Digital State

Digital State

The Story of Minnesota's Computing Industry

THOMAS J. MISA

UNIVERSITY OF MINNESOTA PRESS
MINNEAPOLIS • LONDON

Unless otherwise credited, all photographs and illustrations are reproduced courtesy of the Charles Babbage Institute, University of Minnesota.

Copyright 2013 by the Regents of the University of Minnesota

All rights reserved. No part of this publication may be reproduced, stored in a retrieval system, or transmitted, in any form or by any means, electronic, mechanical, photocopying, recording, or otherwise, without the prior written permission of the publisher.

Published by the University of Minnesota Press
111 Third Avenue South, Suite 290
Minneapolis, MN 55401-2520
http://www.upress.umn.edu

Library of Congress Cataloging-in-Publication Data
Misa, Thomas J.
Digital state : the story of Minnesota's computing industry / Thomas J. Misa.
Includes bibliographical references and index.
ISBN 978-0-8166-8331-4 (hc)
ISBN 978-0-8166-8332-1 (pb)
1. Computer industry—Minnesota—History. 2. Information technology—Minnesota—History. I. Title.
HD9696.2.U25M565 2013
338.4′700409776—dc23
2013023773

Printed in the United States of America on acid-free paper

The University of Minnesota is an equal-opportunity educator and employer.

20 19 18 17 16 15 14 13   10 9 8 7 6 5 4 3 2 1

*To the memory of Erwin Tomash (1921–2012),
a native son of Minnesota, one of the pioneers
with the Engineering Research Associates,
a notable computer industry leader, and the
guiding spirit behind the Charles Babbage Institute.*

Contents

Preface	ix
Abbreviations	xiii
Introduction: Minnesota Goes High-Tech	1
1. Philadelphia Story: Wartime Origins of Minnesota Computing	17
2. St. Paul Start-up: Engineering Research Associates Builds a Pioneering Computer	45
3. Corporate Computing: Univac Creates a High-Tech Minnesota Industry	71
4. Innovation Machine: Control Data's Supercomputers, Services, and Social Vision	99
5. First Computer: Honeywell, Partnerships, and the Politics of Patents	135
6. Big Blue: Manufacturing and Innovation at IBM Rochester	163
7. Industrial Dynamics: Minnesota Embraces the Information Economy	189
8. High-Technology Innovation: Medical Devices and Beyond	219
Appendix: Employment in Minnesota Computing, 1980–2011	233
Notes	245
Bibliography	275
Index	289

Preface

Minnesota's computing industry has such an obviously important history that it may seem odd that it has not been told before. The history of computing, a newly emerging historical field, took form when the "small band of digital pioneers" active originally in the 1940s and 1950s began a spirited debate over the priority claims of who built the first computer.[1] Now there is a small bookshelf of works on the pioneering computer efforts in Philadelphia, Pennsylvania, and Princeton, New Jersey, owing in no small measure to the historical resources that were intentionally made available by the U.S. Army, the University of Pennsylvania, and the Sperry Rand Univac company. The latter was even a prominent sponsor of a landmark exhibit on computing mounted by the Smithsonian Institution. Not surprisingly, these accounts directed a bright spotlight on Philadelphia.[2]

I first considered this story in 2006 when I came to the University of Minnesota as director of the Charles Babbage Institute, a leading research and archiving center that helped create the professional field of computing history since its founding in 1978 in Palo Alto, California. CBI was founded by a group of historically minded computer pioneers led by Erwin Tomash, who figures in these pages as an author as well as employee of the Engineering Research Associates (ERA). Tomash played a key role not least by talking up the burgeoning Minnesota computer scene in the Southern California aerospace industry, a vast early market for computers.[3] At a certain moment in the halls of the Convair aerospace division of General Dynamics, these notable people were at work on Minnesota computers in California: Tomash, the future founder of the computer peripheral maker Dataproducts and later of CBI as well; Marvin Stein, the future founder of computer science at the University of Minnesota; Robert Price, the future president, chairman, and CEO of Control Data; and at least one other executive later prominent at Control Data. They might all have eaten lunch together. Price shared with me his recollections of lunchtime banter as well as the more fundamental lessons about personal relationships and corporate strategy.

When I came to Minnesota, everyone told me how vital and important the state's computing industry had been. Yet, at the time, the existing historical literature had little to say. Perhaps the best readily accessible source was an article on the early history of ERA written by Erwin Tomash and Arnold Cohen and published in the inaugural 1979 volume of *Annals of the History of Computing*.[4] Arthur Norberg, the founding director of CBI, had just published *Computers and Commerce: A Study of Technology and Management at Eckert-Mauchly Computer Company, Engineering Research Associates, and Remington Rand, 1946–1957*, and he and CBI associate director Jeffrey Yost were putting the final touches on *IBM Rochester: A Half Century of Innovation*. At that time, the ERA story was mentioned in most historical accounts, if at all, with an aside about ERA's link to the Philadelphia story.

A much stronger public profile existed for Control Data, with its famous figures Seymour Cray and Bill Norris. Oddly, though, apart from a single adulatory biography of Norris and a breezy profile of Cray, the historical literature on Control Data was rather thin, although the outline of its story was reasonably well known.[5] I first pieced together the core narrative from Robert Price's *Eye for Innovation*. Later, I had the privilege of conducting with him a detailed twenty-hour oral history spanning his remarkable career at Control Data, now a published book.[6] Price once considered writing a book-length history of Control Data but despaired when he recognized that the effort might take him a decade. A complete account of this pivotal company is still sorely needed. Honeywell and Sperry Univac were mostly covered, if at all, in scattered specialist publications.

I resolved to plunge headlong into this captivating tale. At CBI during the academic year 2007–8 we organized a yearlong series of public lectures on the theme of Minnesota's hidden history in computing. The research for those lectures brought me closer to the immense resources of the Charles Babbage Institute; in its absence, this project could never have been undertaken. As I subsequently worked on turning those public talks into these chapters, I delved further into the research materials that were at my fingertips at CBI. These archival riches depend on the professional archivists who have worked at CBI, including the founding archivist Bruce Bruemmer, as well as his successors Beth Kaplan, Carrie Seib, Karen Spilman, Susan Hoffman, R. Arvid Nelsen, and Stephanie Crowe. At CBI I have the great fortune to work alongside Jeffrey Yost and Katie Charlet. As the notes for this book make clear, I have extensively drawn on the Control Data corporate records, the papers of William Norris, a number of individual collections from Univac and Control Data employees, the flagship CBI collection on the *Honeywell v. Sperry Rand* lawsuit, the Mark McCahill papers, and the

Marvin Stein papers, as well as additional documentation on the singular Minnesota Educational Computing Consortium.

Another unique CBI resource is the remarkable set of oral histories, both published and unpublished, that shed light on innumerable episodes as well as contribute color and perspective to these pages. Arthur Norberg did a great number of oral histories with ERA technical staff and executives for his book *Computers and Commerce,* including one with Bill Norris that is far and away the most insightful source about this complex figure. Norris was at the center of the wartime efforts in cryptography, the founding and early years of ERA, the wrenching reorganizations of Sperry Rand Univac, and then the founding of Control Data in 1957. He remained chairman and CEO of Control Data Corporation (CDC) for nearly three decades until relinquishing that executive position to Bob Price in 1986. Many of these oral histories are publicly available on CBI's Web site, but I also carefully consulted dozens of oral histories conducted by Control Data in the early 1980s. These informally transcribed interviews, containing many colorful misspellings and quirky mistakes, helped me understand the deeper culture of the company, because the voices are those of operating managers, division heads, design and manufacturing engineers, and even several early founding figures in the company who tangled with Norris and left for other prospects.

Telephone conversations with Walter Anderson, one of the ERA pioneers, provided valuable personal reflections on the distinctive time and culture of the early ERA years. Other discussions with numerous technical people who worked in Minnesota's computer industry—notably a cluster of Lockheed Martin and Unisys employees and retirees—also connected me back to the ERA days. In addition to my lengthy oral history with Bob Price of Control Data, I conducted a more focused interview with Curt Mathiowetz and Steve Lewis, the lead managers of IBM Rochester's Blue Gene supercomputer project. Mike Svendsen, who was involved with Univac's quality assurance in the 1960s through the early 1980s, let me examine documents in his possession that provide an industry-wide assessment of semiconductor quality.

Remarkably enough, this book is the first examination of the complete dimensions of this notable Minnesota story. *Digital State* tells the story of the emergence of a high-tech industrial district in Minnesota and presents essential chapters in the birth of the digital era.

Abbreviations

ACH	Automated Clearing House
ARDC	American Research and Development Corporation
ARPA	Advanced Research Projects Agency
ARPANET	Advanced Research Projects Agency Network
ARTS	Automated Radar Tracking System
ATC	air traffic control
BUSHIPS	Bureau of Ships (U.S. Navy)
CBI	Charles Babbage Institute
CDC	Control Data Corporation
CDI	Control Data Institute
CEIR	C-E-I-R, Inc., originally the Council for Economic and Industrial Research
Cognac	Cogitating Numerical Adder and Computer
CPU	central processing unit (computer)
CSAW	Communications Supplemental Activity–Washington
C-T-R	Computing Tabulating and Recording Company
DEC	Digital Equipment Corporation
EDSAC	Electronic Delay Storage Automatic Calculator
EDVAC	Electronic Discrete Variable Automatic Computer
ENIAC	Electronic Numerical Integrator and Computer

ERA	Engineering Research Associates
FAA	Federal Aviation Administration (Federal Aviation Agency, 1958–67)
FDA	Food and Drug Administration
FORTRAN	*Formula Trans*lation (IBM programming language)
GE	General Electric Corporation
HP	Hewlett-Packard Corporation
IBM	International Business Machines
ICBM	intercontinental ballistic missile
IEEE	Institute of Electrical and Electronics Engineers
ISO	International Organization for Standardization
LARC	Livermore Advanced Research Computer
LLNL	Lawrence Livermore National Laboratory
3M	Minnesota Mining and Manufacturing Company
MBA	master of business administration
MDM	Minnesota Directory of Manufacturers
MECC	Minnesota Educational Computing Consortium
MICE	Midwest Internet Cooperative Exchange
MIT	Massachusetts Institute of Technology
MMPI	Minnesota Multiphasic Personality Inventory
MTBF	mean time between failures
MTDC	Midwest Technical Development Corporation
NAC	Northwestern Aeronautical Corporation
NASA	National Aeronautics and Space Administration
NBS	National Bureau of Standards
NCML	Naval Computing Machine Laboratory
NCR	National Cash Register Company
NEL	National Electronic Laboratories
NSA	National Security Agency
NSF	National Science Foundation
NTDS	Naval Tactical Data System
OEM	original equipment manufacturer
ONR	Office of Naval Research
OS	operating system

PDP	Programmed Data Processor (DEC minicomputer of the 1960s)
PIRAZ	Positive Identification Radar Advisory Zone (NTDS)
PLATO	Programmed Logic for Automated Teaching Operations (CDC)
RCA	Radio Corporation of America
RISC	reduced instruction set computing
ROM	read-only memory
RPG	Report Program Generator (IBM programming language)
SAGE	Semi-Automatic Ground Environment
SCF	Semiconductor Control Facility (Univac)
SCOPE	Supervisory Control of Program Execution (Control Data)
SEAC	Standards Eastern Automatic Computer (NBS)
SMD	Storage Module Drive (Control Data)
SWAC	Standards Western Automatic Computer (NBS)
UNIVAC	UNIVersal Automatic Computer
USPTO	United States Patent and Trademark Office
VAX	Virtual Address eXtension (DEC minicomputer)
WAVES	Women Accepted for Volunteer Emergency Service (U.S. Navy)
WCN	William C. Norris
WWW	World Wide Web

INTRODUCTION
Minnesota Goes High-Tech

Minnesota's computer industry transformed the state's economy and identity in the years following the Second World War. Part of the reason this story has never been told is that Minnesota's computing companies, while world famous in their own way, were never in the business of selling computers directly to individual consumers. They lacked the brand appeal of "Intel Inside," the pizzazz of a Microsoft product launch, or the buzz of Apple's distinctive products and advertising. Instead, the Minnesota computer companies made their reputations in selling computers to the government or to other industries. Indeed, for three decades or more they built their most highly regarded computers for the military and intelligence agencies, frequently behind the closed door of top-secret classified contracts. It wasn't a secret to everyone. Insiders have known all along that the rise of the computer industry in Minnesota—centered on the Twin Cities and Rochester—worked a dramatic transformation in the state's image and identity. By 1983 the state, despite its legacy of Scandinavian reserve, proclaimed itself as the "Supercomputer Capital of the World."[1]

A world-class computer industry was all the more remarkable considering the state's recent history. Minnesota emerged somewhat late as an organized territory of the United States owing to its geographic isolation as well as disquieting shifts, to white settlers at least, in the Native American tribes.[2] Felling the state's rich bounty of white pine trees in the north and planting the state's rich soils in the south and west became the dominant economic activities after statehood was achieved in 1858, at least until a mammoth bonanza of iron ore deposits was commercialized in the north-central part of the state by the Rockefeller and Carnegie concerns beginning in the 1880s. Railroading and city building in Duluth, St. Paul, Minneapolis, and many smaller places absorbed additional energies in the following decades. In 1922, author

Sinclair Lewis, a native of Sauk Centre, Minnesota, described his prototypical Midwestern city of Zenith (for which both Duluth and Minneapolis have claimed to be the model) as "a city with gigantic power—gigantic buildings, gigantic machines, gigantic transportation. . . . It is one big railroad station."[3]

The state's economy by 1945 was best known for milling flour, weaving woolen textiles, shipping iron ore to other states, and turning cows into beef in the vast South St. Paul slaughter yard. At the time, Minnesota's economy could best be described as agriculture-driven rather than industry- or technology-driven. A distinguished visitor to the land-grant University of Minnesota had even scolded it for an undue focus on the agricultural interests in the state rather than devoting proper attention to its emerging industries. There were indeed some heavyweights in agriculture. General Mills, Pillsbury, Cargill, Green Giant, and Hormel were among the nationally famous Minnesota-based agro-industrial enterprises that powered the state's economy and impressed on it a distinct agro-industrial identity. Even a number of the state's midsized industrial enterprises then developing, such as Lawn-Boy and Toro, two pioneer manufacturers of lawn mowers; Minneapolis–Moline, the prominent farm tractor and agricultural machinery maker; and the toy truck manufacturer Tonka (named with a Dakota Sioux word), had obvious though diverse ties to the soil. So did the

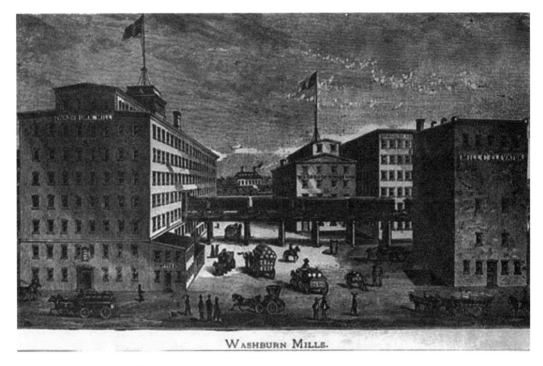

FIGURE I.1. Washburn–Crosby Mills in Minneapolis (here circa 1879) merged with two dozen other firms to create General Mills in 1928. Today these mills are the site for the Mill City Museum along the Mississippi River. Source: University of Minnesota Archives.

new firms Polaris and Arctic Cat, the leading early snowmobile manufacturers that transformed the land into an object of consumption and recreation.

Much of the state's high-technology industry can be traced to the pioneering computer company Engineering Research Associates. Founded in 1946, ERA delivered some of the earliest general-purpose computers in the country, using its wartime expertise in cryptography to help create the new world of digital computing. Although others have established their claims for building the "first stored program computer," as discussed in chapter 2, ERA's main claim to fame was in building and then successfully *moving* such state-of-the-art computers more than a thousand miles—at a time when other early computers typically never left the laboratories they were constructed in. ERA's pioneering computer was designed, constructed, and tested in St. Paul then shipped by rail to Washington, D.C., and doing useful work by December 1950. It is worth noting that although British computer projects beat the Americans to a stored-program computer by a year or two, the American stored-program computer with the strongest priority claim—the National Bureau of Standard's SEAC—was itself being tested in the spring of 1950. It is by no means clear that SEAC's testing regime was comparable in technical rigor to the final testing and refinement of the ERA machine also done during these months. (Chapter 2 evaluates other rival claims, including that for BINAC, which to all appearances was wrecked when it was shipped from Philadelphia to California in the fall of 1949.) ERA also served as a fertile training ground for a distinctive group of far-seeing business entrepreneurs, technically minded scientists, and talented electrical engineers who went on to build several generations of computers for the Univac company as well as create over time perhaps as many as fifty start-up companies.[4]

By 1959, the secret of Minnesota's high-tech industry was out. In April that year the *Minneapolis Star* ran a full-page spread on the "Twin Cities Boom" in electronics, anchored by the giant established firms of Minneapolis–Honeywell, Remington Rand Univac, and Minnesota Mining and Manufacturing (3M). Even General Mills had recently "plunged into electronics hardware production." Far more than an agro-industrial giant and originator of the idealized images of the Betty Crocker housewife, General Mills had made military electronics, torpedo directors, and bombsights during and after World War II. Then, in the mid-1950s, it established a digital computer laboratory on East Hennepin Avenue, collaborated with the University of Minnesota on high-altitude balloon physics, and, at its manufacturing facilities at 1620 Central Avenue in northeast Minneapolis, manufactured customized computers for NASA, the army, and the intelligence agencies. The electronics industry was clearly an economic powerhouse, with direct employment of twenty-five thousand people in the Twin Cities, annual payrolls of $120 million, and sales volume estimated to be upwards of $400 million. Honeywell alone accounted for fourteen thousand workers at its twenty-one plants with an annual payroll of $70 million.[5]

FIGURE I.2. Women programmers at the Minnesota State Fair in 1969. "Program your own future in the computer industry," advised Control Data. Expanding rapidly, the company tapped Minnesota women for its programming workforce.

Many of the classic attributes of a dynamic high-tech economy were already in place. In addition to the established anchor firms, there were no fewer than forty-seven smaller firms, organized mostly in the preceding decade. Many of these were "spin-offs" from the larger firms such as Honeywell (which spun off Cedar Engineering and three other firms), Univac (Control Data, Ramsey Engineering, and at least five other firms), and the University of Minnesota (Rosemount Engineering). Spin-offs, the paper noted, typically resulted when "a minor executive or technician of a major company has recognized a potential market for a specialized service or product and has had the aggressive nature to pursue the possibility." Here, already realized in 1959, was a formula for success followed across the country and around the world. Another spin-off from Univac and Control Data was Midwest Technical Development Corporation, the country's second venture capital firm (see chapter 4). In funding the region's electronics boom, Midwest Tech was joined by three recently formed small-business loan firms as well as by the established First National Bank, whose vice president commented favorably that "Interest in electronics investment is growing tremendously, both by the public and by banks."[6]

While Control Data's original one-dollar stock was already worth sixteen dollars a share at the time of the *Minneapolis Star* profile, within two additional years it jumped to forty dollars a share and began a meteoric rise to over one hundred dollars. In November 1961, the *Upper Midwest Investor* published a special issue profiling the Twin Cities computer industry. Control Data now warranted a full write-up alongside Univac, General Mills, and Honeywell, where (it was said) "computers were a natural." The story was of national importance. Manufacturers of computers, peripheral equipment, displays, and circuitry, along with the services companies that rent computer time, "are grouping themselves around a certain few metropolitan areas," the article noted. "One of these is that of the Twin Cities." The industry employed eight thousand people directly in the design and manufacture of computers, with additional thousands employed in making the components, peripherals, and mechanical parts needed for computers. It was estimated that Univac alone spent around $70 million a year in the region (about half on payroll) while IBM in Rochester spent $35 million.

Then, as now, the question posed by the *Upper Midwest Investor* was why this impressive computer boom—in Minnesota? The author at the time pointed to the state's "excellent" educational facilities, from grade schools through "big state universities," the "relatively temperate" weather, the readily available industrial sites with good transportation, and the willing availability of public financing through local banks and investors. An "excellent labor supply" and a solid work ethic substantially outweighed the state's tax structure, according to all the computer company officers that were interviewed. Pointing to the sizzling stock performance of Control Data, the author noted that "the high investor interest in technical companies . . . played a large part in the establishment of the computer industry in the state." It so happened that the attractions of Minnesota's weather (especially the pleasant summers) lost out over the years with the rise of air conditioning.[7]

These two popular surveys in 1959 and 1961 were visible signals of Minnesota's emerging high-technology industry. It is little recognized today that the state's computer industry—the state's flagship companies besides ERA included Sperry Rand Univac, Control Data, Honeywell, and IBM Rochester, each depicted in the following chapters—was for at least two decades the largest and most varied in the United States. Even though the region was eventually surpassed by IBM's vast manufacturing complex in upstate New York, the minicomputer industry in Massachusetts, and the software and personal computer industry of California's Silicon Valley, Minnesota computing remained of impressive size and diversity for years. The state's computer industry, drawing on graduates and researchers from the University of Minnesota, created sixty-eight thousand computing jobs in the state by 1989. This book tells the story of building a computer-centered high-technology economy. It combined entrepreneurial vision, technical genius, advanced know-how, steady financial backing, and judicious government support. This is how Minnesota became a digital state.

6 INTRODUCTION

FIGURE I.3. Vote totals were tabulated by Control Data computers on Election Night 1982 at Minneapolis WCCO–TV News. From left: Skip Loescher, Dave Moore, Pat Miles.

Minnesota, we can now appreciate, created an innovative computer-centered *industrial district,* one of the very first of its kind. Industrial districts have fascinated many observers ever since the rise of Boston's Route 128 and California's Silicon Valley, as well as Bangalore, India, and the hyperspecialized industrial districts in southern coastal China that make, sell, and ship the world's supply of socks, bras, and neckties.[8] The concept has recently gained traction in the social and historical analysis of economic development. These accounts begin with the "classic" industrial districts in Britain that Alfred Marshall, in his *Principles of Economics* (1890), had in mind when he coined the term. The particularities of place, skill, history, and innovation were much on his mind:

> When an industry has thus chosen a locality for itself, it is likely to stay there long: so great are the advantages which people following the same skilled trade get from near neighbourhood to one another. The mysteries of the trade become no mysteries; but are as it were in the air, and children learn many of them unconsciously. Good work is rightly appreciated, inventions and improvements in machinery, in processes and the general organization of the business have their merits promptly discussed: if one man starts a new idea, it is taken up by others and combined with suggestions of their own;

and thus it becomes the source of further new ideas. And presently subsidiary trades grow up in the neighbourhood, supplying it with implements and materials, organizing its traffic, and in many ways conducing to the economy of its material.⁹

Economists and historians have found that the industrial-district concept facilitates understanding of specialty manufacturing in Italy, France, the United States, and many other countries. Regional planners, city officials, business leaders, and others are seeking to enhance their own economic development. Philip Scranton, in his masterful survey of U.S. specialty manufacturing, *Endless Novelty*, suggests a multipart typology to capture the specific relations among firms.¹⁰

Three of Scranton's core concepts, plus a bit of extension to account for globalization, result in a robust framework to analyze a high-tech industrial district. This book's core chapters profile the *"integrated anchor"* firms in computing. Notably in Minnesota, such large companies as Univac, Control Data, Honeywell, and IBM were able to assemble the range of financing, technology, and skills needed to design and manufacture digital electronic computers. Univac and IBM were local divisions of internationally prominent corporations, while Control Data and Honeywell grew from local origins into nationally prominent enterprises in their own right. Because these firms played a prominent role in government and industry, not least by attracting the attention of national capital markets, such as the New York Stock Exchange, they loom large in the public mind. These core chapters relate how the Minnesota-based managers and engineers at Univac and IBM responded to the opportunities afforded to them as divisions of large corporations as well as how Honeywell and Control Data, each in distinct ways, grew from local origins into positions of unusual prominence and national renown. The wide-ranging activities of these companies created a notable high-tech economy and a distinctive identity for Minnesota as a digital state.

The anchor firms were so prominent that the wider ecosystem consisting of smaller and larger firms and supporting institutions tended to slip out of focus. Here, Scranton's notion of *"specialist auxiliary"* firms spotlights an essential, if not always readily visible, dimension of industrial districts. Many specialist auxiliaries manufactured components or subcomponents needed in the computer industry. Mostly, the integrated anchors were content to rely on networks of smaller firms for such necessary components as wiring harnesses, printed circuit boards, sheet-metal assemblies, heat sinks, and other specialized subassemblies. Other specialist auxiliaries vital to the computer industry include high-precision machine shops as well as firms dedicated to design, engineering, and prototyping. When Control Data built its own factories for high-volume manufacturing of tape drives, memory units, and other computer peripherals, it was to some extent taking up the function of auxiliaries that were manufacturing these very same items. (In Scranton's original typology, one could say that Control Data became a "bridge firm" blending both specialist and commodity production formats.) Not all specialist auxiliaries remained small. Founded in 1965 as a small concern, Hutchinson Technology employed three thousand Minnesotans in 2011, focusing on a hyperspecialized subassembly needed in magnetic disk drives. Lawson Software, founded in 1975, drew on the programming workforce assembled by the anchor firms to become one of the leading producers of enterprise software, and employed five thousand Minnesotans in 2011.

A third class of firm forms the *"ancillary industries"* that round out the state's industrial district. These firms create a wider supportive ecosystem for the high-tech companies. The giant Minnesota Mining and Manufacturing Company (3M) played a direct role in computing through interactions with the pioneering ERA company in magnetic coatings, while a Honeywell executive later led Memorex, the well-known maker of magnetic media. One might even suggest that the financial services industry forms something of an ancillary to computing. At the very least, the Twin Cities' multitudinous commercial banks, credit unions, mortgage brokers, real-estate firms, insurance companies, securities dealers—and a half dozen or so venture capital firms—employed more than ninety-eight thousand Minnesotans (1992 figures) and deployed untold numbers of computers.[11] Chapter 7 touches on the key role of the Minneapolis Federal Reserve in shaping the regional economy and the demand for computing.

Recent authors exploring the distinctive aspects of innovative regions in computing include Christophe Lécuyer and others on Silicon Valley and Paul Ceruzzi on the Internet Alley just outside Washington, D.C.[12] Each of these authors is going beyond the early enthusiasms for "high tech" and, like this present work, seeking to understand how specific places shape technological innovation and economic change. This book's focus on "industrial districts" also highlights and clarifies important continuities between the state's computing industry and the medical-device industry that followed in its wake. I examine the lessons for strategizing about state-level economic development in chapter 8.

The history of computing today naturally inclines to the business entrepreneurs, engineers, and scientists who launched such world-famous companies as Hewlett-Packard, Intel, Apple, and Google and made Silicon Valley into a global technology juggernaut. No one thinks of this as a merely local history of some California companies, even though it's essential to recall that "Silicon Valley" did not exist until a business journalist coined the term in 1971. Or, to take an earlier period focusing on the invention of digital computing in the 1940s, you might instead zero in on developments across the Atlantic in Manchester, Cambridge, or Berlin, as well as in Philadelphia and Princeton; this earlier history, too, related events necessary to comprehend the emergence of modern computing. The earlier history might link the names of Alan Turing, John von Neumann, Claude Shannon, and other computing pioneers with the universities, government agencies, and companies that created modern digital computing in the closing days of the Second World War and the years that followed.

This book tells an intermediate story: the history that lies between the founding moment of digital computing in the years around 1945 and the ascendency of Silicon Valley in the 1970s. While this history has a significant relation to the Philadelphia origin story, its true beginnings are the tangle of high-technology intelligence operations surrounding the attack on Pearl Harbor and the massive government mobilization of digital electronics and computing. The link between intelligence and technology, while novel, was not at all obscure. During the Second World War, governments on both sides of the conflict mobilized scientists and engineers like never before to invent and innovate new technologies, and these were rushed into factories and often enough onto the battlefield. While Germany took the lead in mechanical and aeronautical technologies, launching rockets that terrorized Britain (while subsequently providing both the Americans and the Soviets with expertise and personnel critical for the space race) and even flying a prototype jet fighter before the war was over, the British and Americans devoted immense resources to atomic energy and electronics.[13] In several instances, including the atomic bomb, the British and Americans established cooperative joint efforts that put into play such technology wonders as high-frequency radar, proximity fuses, and a bevy of top-secret code-breaking machines that should properly be seen as early digital computers.[14]

Within a year of the atomic bombs falling on Japan, as the United States demobilized its soldiers, sailors, and scientists, the unique proto-computing expertise built up during the war years in Dayton, Ohio, and Washington, D.C., was transferred across the board to St. Paul, Minnesota. Beginning with the founding of the Engineering Research Associates in 1946 as a distinctive blend of a contract-driven innovation laboratory and military think tank, Minnesota developed unique technical expertise in cryptography and digital electronic computing.

These seminal developments in electronic digital computing have not received the attention they deserve, owing a great deal to the "top secret" classification that hindered public awareness of these Minnesota-based developments. The Minnesota computing machines remained shadowed in secrecy for many years after the U.S. Army made a showy public display of its ENIAC calculating engine in 1946. Other postwar circumstances that shaped the public's awareness of the origins of digital computing, besides the army's ongoing publicity, include the Philadelphia-centered effort to build early computers for commercial sale that culminated with the Univac computer and the controversial bid by famed Princeton-based mathematician and early computer theorist John von Neumann to collar the claims for conceptual advances in computing. It was customary for other computing pioneers to closely watch their tongues when speaking about secret work and government projects, especially during the early Cold War years, the period of Senator Joseph McCarthy's red-baiting, and the tense years of atomic brinksmanship and pervasive anxiety about state secrets. Such concerns scarcely bothered von Neumann. He was connected to the very highest levels of military and atomic security, as chapter 1 discusses in additional detail, and his word to all intents and purposes was law.[15]

ERA was a pathfinder in yet another way. ERA's technical successes were especially well known in the top-secret world of military and national-security computing, beginning a long tradition of Minnesota companies that were famous in Washington, D.C., and around the world in military and government circles but whose classified computer systems—as well as the development, manufacturing, maintenance, and support activities they depended on—could not be publicly acknowledged and celebrated back home. "Many of the dramatic successes in terms of capability to perform a mission as well as to generate revenue and profitability, many of those early successes were not publicized because they tended to be for customers who were involved either in intelligence work like the National Security Agency, or in defense department activities of a classified nature," stated one prominent ERA–Univac manager.[16]

It's difficult to imagine, say, Seattle with Microsoft unable to shout out its successes, or Silicon Valley with such firms as Intel, Hewlett-Packard, and Apple confined to fame only in the government world, but that is what existed—for decades—in Minnesota. In addition to ERA's supersecret cryptographic work, Univac made computers that networked the U.S. Navy's battle fleet and that landed commercial and military airplanes across the country and around the world. Control Data sold many of its top-of-the-line supercomputers to the nation's nuclear weapons laboratories and supersecret intelligence agencies. Honeywell, the state's largest private-sector employer for many years, was such a prominent military contractor that it attracted a nationwide antimilitary protest movement, as related in chapter 5. Even IBM's Rochester-designed and -manufactured Blue Gene supercomputer was commissioned by the Lawrence Livermore National Laboratory for its mission in nuclear simulation and stewardship. Especially in the southern suburbs of the Twin Cities, there were for years specialized military computer divisions of Lockheed-Martin, General Dynamics, and other military behemoths.

It is important to recognize that the core companies in this history—Engineering Research Associates, Sperry Rand Univac, Control Data, Honeywell, and several others—were once the high-technology bellwethers just as Apple and Google are today. These Minnesota companies figured prominently in the dossiers of military and intelligence officers, but also in the pages of the local newspapers, the screens of the local television stations, and the aisles of the perennially popular Minnesota State Fair (see Figures I.2. and I.3.). Every once in a while, they made it to the screens of Hollywood (Figure I.4.). A select few of the people behind the Minnesota computer companies, such as Seymour Cray and Bill Norris, are famous even today to

FIGURE I.4. *Colossus: The Forbin Project* was a 1970s science-fiction thriller about an out-of-control supercomputer. Control Data gave equipment worth $4.8 million to the movie's producers, aiming for favorable publicity.

computer aficionados. At their peak of influence and activity, each of these companies was rooted in the state of Minnesota, again anticipating the dimension of regional development that figures in the Silicon Valley epic.

To provide an overview of the chapters that follow, we can briefly highlight six notable developments in computing that this history places at the center of its narrative. To begin, this book makes a compelling case that the first distinctive computing industry took form in the Twin Cities. While other cities with significant early computing activity such as Philadelphia, Detroit, Dayton, Boston, and Endicott or Poughkeepsie in upstate New York were typically dominated by one or two large computing companies, in the Twin Cities an early pattern of "spin-offs" and "start-ups" created by the early 1960s at least thirty computing companies active in the state. Some of these, such as Control Data, founded by former Sperry Rand Univac managers and engineers, became world famous in their own right, while dozens of lesser-known firms were well regarded at the time, added significant economic activity and technical expertise to the region, and sometimes were merged into larger entities. It was the co-located presence of several prominent anchor firms, including Univac and Control Data, that created a pioneering industrial district specializing in computing.[17]

Also moving decisively into digital computing around 1960, and adding significantly to the economic volume and technical sophistication of the state's computing industry, were Minneapolis Honeywell and IBM Rochester. IBM's expanding factory complex in Rochester, Minnesota, soon developed not only expertise in high-technology manufacturing but also significant prowess in computing innovation. Rochester's engineers and managers designed, developed, and manufactured more than four hundred thousand units of a massively lucrative midsize computer—and this was no personal computer because its selling price ranged from $200,000 to more than $1 million—generating larger revenues in 1990 than those of the entire Digital Equipment Corporation, far and away the leading minicomputing company and at that time the number two computer company. It sounds peculiar, but one could say that even as IBM was the world's largest computer company, IBM Rochester was something like the world's second-largest computer company. Minneapolis Honeywell was for many years the state's largest private-sector employer; the large payrolls also at Control Data, Univac, and IBM Rochester added additional thousands to the state's computing employment, which in 1989 stood at sixty-eight thousand and accounted for 3.3 percent of the state's entire workforce—a proportion of computing employment exceeded only by Massachusetts, then at the peak of its minicomputing industry.

A sign of the prototypical industrial district in Minnesota is that its computer engineers experienced stock ownership, informal hiring networks, and extensive intercompany mobility decades before these features were supposedly "discovered" in the Boston metropolitan area's Route 128 district and California's Silicon Valley.[18] The Twin Cities had its local restaurants where engineers from different companies informally swapped technical tips and employment prospects long before Silicon Valley's Wagon Wheel restaurant made these common practices famous.[19] It is also notable that Minnesota had the nation's second venture capital firm, Midwest Technical Development Corporation, founded in 1958–59 and modeled explicitly after Boston-based American Research and Development Corporation.

A second persistent characteristic in this history is Minnesota companies' subtle use of powerful magnets. While the earliest computers relied on troublesome vacuum tubes and finicky acoustic schemes to store data and instructions, Minnesota companies took an early lead in developing and commercializing magnetic media—the basic technology at the center of all hard-drive storage and the physical location of today's cloud computing—through mobilizing expertise from the University of Minnesota and a partnership with the regional giant 3M, the Minnesota Mining and Manufacturing Company. Well known for its sandpaper and industrial adhesives, 3M, located in St. Paul since 1910, invented a spray-on magnetic coating that was tested and refined by ERA as it developed its own magnetic storage devices, an early collaboration that helped 3M achieve global hegemony in magnetic data and sound recording. ERA, at the time struggling for financial stability, took up a contract with IBM to develop magnetic technology, which resulted in the transfer to IBM of "two massive patents" that provided crucial intellectual property and technical insight in the newly emerging field. IBM capitalized on this custom development work, and added its own significant manufacturing and marketing savvy, to build the first mass-produced computer. For storage the IBM model 650 utilized a magnetic drum that was instantly recognizable to the ERA engineers. While ERA sold dozens of its high-end computers with magnetic drum storage, IBM sold a phenomenal two thousand units of the modestly priced model 650 from 1954 to 1962.

The physics and technology of magnetic storage remained a special area of expertise in Minnesota. Control Data built a lucrative third-party or OEM business in large measure by mass-manufacturing and then selling its magnetic storage units under other companies' branding. When Control Data spun off its magnetic storage division in the late 1980s, it was snapped up by Seagate, which used the acquisition to bolster its position in the high-end market for magnetic storage and consequently

FIGURE I.5. The Minnesota Supercomputing Institute (established in 1986) involved a complex partnership between the state and the university that aimed to develop an urban high-tech district along Washington Avenue in Minneapolis.

emerge as one of the principal players in today's highly competitive disk-storage industry. Imation, the notable magnetic media company, was a spin-off from 3M. And even today Hutchinson Technology, one of the specialist auxiliaries employing three thousand in Minnesota, focuses on making a specialized internal subassembly necessary for magnetic disk-drive memory units.

A third area of technology frequently highlighted in these pages is the perennially exciting category of supercomputing. Supercomputing is indelibly attached to Minnesota owing to the Control Data Corporation and especially its charismatic Seymour Cray. Cray's design for the CDC model 1604, an early commercially successful computer based on transistors rather than vacuum tubes, put Control Data squarely into the computer business, while his subsequent CDC model 6600 broke the mold. It was not merely three times faster than the IBM 7030 Stretch that it displaced as the world's fastest computer but also was manufactured in quantity and essentially defined the term "supercomputing." Estimates vary, but at least one hundred of these $5-plus million machines were sold. For twenty years, Control Data essentially owned the supercomputer business of the pace-setting Lawrence Livermore National Laboratory. From 1963 to 1982, Lawrence Livermore exclusively purchased top-of-the-line Control Data or Cray computers, with the exception of two IBM 7094s. Additional CDC machines went to Livermore's rival Los Alamos National Laboratory and other government supercomputing facilities, including the National Security Agency. "It was almost a tradition that one of the first of any new faster CDC machine was delivered to a 'good customer'—picked up at the factory by an anonymous truck, and never heard from again," stated one Control Data engineer, with a broad wink aimed at the NSA. The coveted title of world's fastest computer returned to the state with IBM's Blue Gene/L that was designed, engineered, and manufactured in Rochester, Minnesota. Fittingly enough, the inaugural copy was destined for Lawrence Livermore. While Rochester's Blue Gene was rated as the world's fastest supercomputer for several years, as described in chapter 5, yet another Rochester-built computer powered IBM's more recent tour on the popular quiz show *Jeopardy!* with its Watson artificial-intelligence machine.

In Minnesota, no less than anywhere else, the legal environment defined and shaped the evolution of the computer industry. In several instances, lawsuits in Minnesota had national repercussions. A Minnesota judge in October 1973 handed down a legal decision that decreed who "invented the first computer" and seemingly settled a patent-law dispute that had bedeviled the industry since the ENIAC patent. The epic case of *Honeywell v. Sperry Rand* was the largest and longest federal law case at the time.[20] Although the trial itself took two years, from 1971 to 1973, the legal skirmishing had been going on since the late 1940s. It is quite a tale itself to explain why the ENIAC patent was not issued until 1963, nearly twenty years after the project began in the middle of the second world war. The legal battle resulted in a titanic volume of invaluable documentation on the early computer industry, with the trial transcript eventually reaching more than twenty thousand pages. There was one further sign of the times. Honeywell's legal team, eventually victorious, extensively utilized a novel computer database to keep tabs on nearly thirty thousand pieces of evidence that were submitted in the trial. Even apart from its own substantial achievements in computing, Honeywell may be the Minnesota company in these pages that is longest remembered. The Minnesota Historical Society named the Honeywell "round" thermostat as well as the antiwar Honeywell Project to be among the 150 most notable people, places, and events that shaped the state's history.[21]

Moving closer to the present, this history explores the origins of "computer services" that today guide the business strategies for such global technology heavyweights as IBM, Xerox, Infosys, Tata, Google, Amazon, and many others. The idea of selling a package of integrated "computer services" and not just hardware or software was conceptualized and prototyped at Control Data, which sought to integrate the company's varied capacities to solve real-world problems for clients. Like Honeywell, Control Data developed an intricate computerized database to successfully wage its own landmark legal battle against IBM (1968–73). One result was the forced bargain-basement sale to CDC of IBM's Service Bureau Corporation. With its proprietary Cybernet computer network, plasma-based Plato work stations, and in-house Service Bureau Corporation, Control Data aimed to deliver innovative educational, weather, financial, and legal services. Control Data was certainly ahead of its time with computer services. One well-placed CDC insider recalls that the financial analysts that were crucial to the company's reputation on Wall Street had distinct indifference to the company's effort to commercialize computer services. Instead, the analysts seemed to be stuck on the glory days of supercomputing, always asking what Seymour Cray was up to. The shift to computer services famously achieved by IBM's Lou Gerstner in the mid-1990s also directly impinges on this history, because this challenging strategy set the environment for IBM Rochester's recent efforts in computer innovation.

And, finally, it is important to appraise the University of Minnesota's role in fostering the state's computer industry. Two of the university's successful technical departments, Electrical Engineering and Computer Science, have had a long history of contributing to the computer industry as well as providing much of the industry's technical talent. It is said that ERA hired fully two-fifths of the electrical engineering graduating class of 1943. Cray himself gained two degrees from the university. He once told an interviewer, "I was fortunate in having an instructor at the University of Minnesota who was looking after me in the sense that when I said 'What's next?', he said 'If I were you I'd just go down the street here to Engineering Research Associates and I'd think you'd like what they're doing there.' This turned out to be one of the very first computing facilities in the sense of developing digital circuits." By the mid-1960s, Cray was among the four Control Data board members with significant ties to the university as alumni, consultants, or donors.

The university's computing center and computer science program were pegged to Sperry Rand Univac's provision of computing resources, as detailed in chapter 2. In effect, the company gave the university extremely valuable computing time with the requirement that it develop advanced courses that treated computing as a topic of "intrinsic interest" and concern. Then there is the University of Minnesota's early version of what became the World Wide Web. Before the World Wide Web came to prominence around 1995, the Internet Gopher was the most popular means for posting, sharing, and searching multimedia content on the open, public Internet. Its inventor, Mark McCahill, working at the university, also coined the phrase "surfing the Internet" to describe the playful yet purposeful exploration of far-flung computer networks (see Figure 7.11). Through the mid-1990s, there was significantly more Internet traffic on the worldwide network of Gopher servers than on the fledgling WWW.

The computer industry profiled in these pages is obviously an integral part of Minnesota's history. It is also a vital part of the state's present-day economy. This book's focus on industrial districts helps clarify the connections between the state's computer industry and its expansive medical-device industry today. These connections between successive high-tech sectors have been insufficiently recognized. Even though the anchor firms in computing have somewhat passed from the scene, the state's computer-based industrial district—including the specialist auxiliaries and ancillary industries—directly supported the rise of medical devices and is very likely the key to future innovation. Chapters 7 and 8 outline the state's diverse and dynamic information economy as well as spotlight the numerous links between high-tech sectors. There is an essential lesson for the state of Minnesota to learn as it ponders its strategies for future economic development. This book is an indispensable place to gain a full understanding of the emergence of Minnesota's high-tech economy and its prospects for the future.

1
Philadelphia Story

Wartime Origins of Minnesota Computing

In its early years, Minnesota's computer industry was improbably but decisively yoked to the Philadelphia story. It's odd that it turned out this way. The chronicles of digital computing often trace its protean beginnings across the Atlantic to Berlin, Manchester, Cambridge, or possibly even to London, if you credit Charles Babbage's mechanical device that dazzled visitors to his fashionable salon in the nineteenth century. By rights the origins of American computing should have been in Dayton, Ohio, and the factories of National Cash Register. But Philadelphia was home to an important wartime computational complex at the University of Pennsylvania. It was also a short train ride from a Maryland army laboratory that had an unusual computing problem as well as deep pockets and military clout. The Philadelphia-based effort to build the ENIAC machine during the war—and commercialize the invention after the war—was one origin for the American computer industry. And, not least, the pioneering Philadelphia computer company and the pioneering St. Paul computer company were joined together when business machine giant Remington Rand bought them both and fitfully merged them, as chapters 2 and 3 will recount. Fallout from the resulting Philadelphia–St. Paul fracas triggered the formation of Control Data, discussed in chapter 4. And the Philadelphia story returned to a Minnesota courtroom in the landmark *Honeywell v. Sperry Rand* lawsuit treated in chapter 5. This chapter locates the origin of Minnesota's computing industry in the military, industrial, and intelligence mobilizations of the second world war.

Washington at War

We now understand that World War II was an information-driven war—in the air, on the ground, and especially at sea—but the secret was unusually well kept for many years.¹ Countless other wartime technical innovations were well known at the time. When two atomic bombs exploded over Japan, hastening the end of the war itself and ushering in the nuclear age that followed, they brought fame to the Manhattan Project and to the army general who led it (see chapter 3), while the effort to develop high-frequency radar for combat helped transform MIT into a research and engineering powerhouse. Proximity explosive fuses, synthetic rubber, operations research, and mass-produced penicillin were acclaimed wartime innovations as well. The massive mobilization of American industry was squarely in the public eye. Thousands of workers churned out unprecedented numbers of ships, airplanes, tanks, jeeps, radios, and a veritable river of gasoline and diesel fuel. There was simply no way to keep the wraps on the floodtide of military supplies issuing from the huge shipyards and automobile factories.

Minnesota's industrial base and educational capacity secured an important role in the nation's military mobilization. The army built the immense Gopher Ordnance Works twenty miles south of the Twin Cities where three thousand workers made gunpowder for artillery shells, while the University of Minnesota trained thousands of technical workers for wartime factories, including more than a hundred women who built airplanes for the Curtiss-Wright Corporation. The ever-present battlefield K-ration was created at the university. "We bought the stuff down at Witt's, the best market in the Twin Cities in those days," physiologist Ancel Keys explained.² War-related work at Honeywell and 3M, as noted in chapters 2 and 5, prepared these companies for the postwar economy. In comparison, the wartime achievements in code breaking that paved the way to the postwar explosion in computing—including the birth of Minnesota's pioneering computer company—were scarcely recognized at the time.

Code making and code breaking—the basic elements of cryptography—were the most closely guarded of all the wartime activities. It's not easy to recall how this was so. For two months during the war, Alan Turing and Claude Shannon had tea together every day in the Bell Telephone Laboratories cafeteria in New York City, but because they were in different research cells they could not breathe a word about their top-secret work. They talked mostly about Turing's remarkable "computable numbers" paper of 1936, which soon inspired a compelling vision of thinking machines and artificial intelligence. Shannon's wartime work created the basis for information theory as a problem in sending and receiving messages over a noisy and unreliable communications channel. "The work on both the mathematical theory of communications and the cryptology went forward concurrently . . . they were so close together you couldn't separate them," he noted.³

Turing's seminal work in breaking the German wartime codes remained entirely out of the public eye until the 1970s, long after his early and tragic death in 1954. At war's end, Winston Churchill ordered that his country's secret code-breaking

FIGURE 1.1. The University of Minnesota trained hundreds of workers for the wartime aircraft industry. Here is a workshop for airplane-propeller mechanics, circa 1942. Source: University of Minnesota Archives.

FIGURE 1.2. Claude Shannon (circa 1950) built his famous "mouse maze" from telephone relays, creating an early learning machine. His wartime work blended telephone switching with cryptography. Source: AT&T Archives.

machines be smashed into "pieces no bigger than a man's hand." His *History of the Second World War* (1948–53) filled up six fat volumes with nearly two million words but revealed, remarkably enough, nothing of his own strategic reliance on a small army of code breakers. Churchill's mandate for postwar secrecy had the effect of creating a hidden history for early British computing, especially compared with the extensive publicity given to the U.S. Army's ENIAC and EDVAC computers.[4] Minnesota's computing industry resulted also from secret wartime work in intelligence and cryptography. Pearl Harbor, in this and much else, was ground zero.

America's intelligence efforts before and after Pearl Harbor suffered from long-standing rivalries between the army and navy. The two services' cooperation in intelligence gathering, such as it was, consisted in taking alternate days to be responsible for analyzing and reporting significant findings to the military command in Washington. Some part of the Pearl Harbor fiasco in 1941 must be traced to the agonizing hours it took to relay the U.S. Navy's fortuitously intercepted message from Tokyo—ordering the Japanese ambassador in Washington to break off diplomatic relations and immediately destroy his cipher machine, effectively a declaration of war—to the U.S. Army chief of staff General George Marshall, the country's top military commander. Early on, American code breakers were routinely reading Japan's diplomatic dispatches. On Saturday, the day before Pearl Harbor, the navy's West Coast listening station began picking up unusual Japanese messages. "As it was an even-numbered date and therefore the Army's duty day," the navy routed the incoming intercepts to the army's Washington-based Special Intelligence Section, and within hours their code machines were printing out the decrypted message in Japanese.[5]

Even more striking were the thirteen short messages sent that fateful Saturday afternoon to the Japanese ambassador in Washington, because they were already in *English* and ready to be delivered to the U.S. government—directly and without any possible errors from translation. Clearly, this was highly unusual. By then, to beef up its thin weekend staffing, the army had secured the navy's assistance with decryption, and so that evening it was navy commander Alwin Kramer who hand-delivered the alarming messages to the select intelligence list, starting with the White House at 9:30 p.m. President Franklin D. Roosevelt read the papers and saw that war was imminent. But something went amiss with the army's side of the decrypt delivery. It seems that army chief Marshall did not get the menacing news until midmorning on Sunday (Washington time), about two hours after a final fourteenth message from Japan directed the ultimatum be delivered at 1:00 p.m. that afternoon, which was around daybreak in the Pacific. With little more than an hour left, Marshall ordered an alert to go out to the Pacific forces: MacArthur in the Philippines seemed the most likely target. Unfortunately, heavy static knocked out the army's radio communication with Hawaii and so the message went via Western Union. It arrived, after a radio hop from San Francisco, in the Honolulu offices of RCA at 1:03 p.m. (Washington time). Lacking any "urgent" label, it was put into a queue for routine delivery to the army's Hawaii headquarters. At the same moment an impossibly huge fleet of airplanes was spotted on the radar screen coming in from the north. Perhaps it was a scheduled incoming flight of B-17s, a duty officer suggested. No one imagined that 183 Japanese warplanes were about to pound Pearl Harbor.[6]

The bungling of communication that led to the "surprise" attack on Pearl Harbor lent special urgency to the U.S. Navy's wartime intelligence operations. The problem was not merely rivalries with the army but also rivalries within the navy between

FIGURE 1.3. Pearl Harbor, December 7, 1941: Battleship Row in the early minutes of the Japanese attack. The "surprise" attack owed something to a tragic slipup in army–navy intelligence. Source: U.S. Navy.

its intelligence and operations branches. Owing to the quirks of history the navy's principal code-breaking unit, the obscurely named Communications Supplemental Activity–Washington, CSAW (see-saw) for short and OP-20-G for long, reported to the Chief of Naval Operations rather than to the navy's intelligence branch. Kramer, relieved of his post after Pearl Harbor, spent the rest of the war testifying to Congress about the varied slipups and misunderstandings between naval operations and naval intelligence, as did his boss Laurance Safford.[7] To straighten out this catastrophic embarrassment, and to prevent another Pearl Harbor, the navy named Joseph Wenger as commanding officer of CSAW.

Wenger was a tall, severe man whom classmates at Annapolis (class of 1923) had nicknamed "Skinny" and "Buzzard." In the 1930s, he learned cryptography at the side of the legendary Agnes Meyer Driscoll. Fluent in five languages, including Japanese, she cracked Japan's blue book, red book, and naval grand maneuver codes (so opening Japan's diplomatic secrets to American eyes), while training the navy's leading cryptographers. Alternating between sea and shore assignments, Wenger already in 1937 had authored a blueprint for "Communication Intelligence Research Activities." He was also an early advocate of using IBM tabulating machines to evaluate the multitudinous letter–cipher frequencies and other clues that might lead to cracking a code itself. To lead CSAW's research branch, Wenger recruited Yale University mathematician Howard Engstrom, "a real gentleman and a real scholar," who was already fluent in German and a largely self-taught code breaker. Under Engstrom was a University of Minnesota math professor, Howard Campaigne, and a Westinghouse sales engineer, Bill Norris, who took charge of a small group that set up the intercept sites. Campaigne, largely behind the scenes, and Norris, highly visible in public, each would play major roles in bringing computing to Minnesota.

After clasping the reins of command at CSAW, Wenger assembled a crack team of mathematicians, radio experts, translators, and steady operators who worked in seventeen technical and support divisions and eventually numbered more than 3,700. Their wartime effort was relocated to a former girls' school located in northwest Washington, D.C., next to American University. Ringed with an imposing double fence of barbed wire, the secure facility became known as the Nebraska Avenue office complex. "Rooms and entrances that cascaded down otherwise hidden stairwells" were among its memorable features. "The maze of narrow corridors . . . seemed like the secret passageways ripped from the pages of some mystery novel; or from a medieval castle."[8] After housing a succession of navy and intelligence offices over the years, in 2003 it became the headquarters of the newly created Department of Homeland Security.[9]

From the buzz and click of faint radiotelegraph signals snared from thin air, Howard Engstrom's unit within CSAW labored to extract information that might shape the course of the war. Engstrom's group dealt "primarily with the intercepted traffic and how to manipulate it and work with it for purposes of preparing it for frequency counts and things like that, where you get repeats of certain information . . . so that you can then deal with the actual assigning of final values to the code groups so that you can deal with them and decide what they really mean."[10] "It was all very exciting actually," one CSAW staffer recalled. "I hate to say it, but the war was one of the high points of my life."[11] Grace Hopper left her teaching post at Vassar and made plans to join Engstrom's operation in Washington (she had completed her mathematics Ph.D. at Yale—the eleventh woman to have done so—in 1934 with one of Engstrom's faculty colleagues) but the navy sent her instead to Harvard, where she spent the war years programming the IBM-built Harvard Mark I under the command of Howard Aiken. In a roundabout way, Aiken's machine played a small role in designing the atomic bomb and a significant role—through Hopper's remarkable career later in Philadelphia—in creating the field of computer programming.

With Japan's diplomatic code pretty much an open book, the tough problem for the CSAW code breakers was the secure code used by the Imperial Japanese Navy. The diplomatic code relied on a simple "additive" scheme that could be decrypted using banks of punch-card tabulating machines (known as "the child of IBM"), but

FIGURE 1.4. Nebraska Avenue office complex in Washington, D.C., in the 1930s, later the wartime site of top-secret code breaking that led to Minnesota's pioneer computer company. Source: U.S. Navy.

the navy code used a far more complex substitution–transposition scheme. When the Japanese Navy unaccountably delayed the changeover of code books in the spring of 1942, and a flood of intercepted radio traffic about fleet movements in the middle Pacific soon followed, "no one needed to be told to go on working long after the rest of the building had emptied."[12] The extra effort resulted in several consequential decrypts. CSAW intelligence was strategically used in the battles of Coral Sea and Midway (May and June 1942, respectively) to make the most of thinly spread U.S. Navy forces. At Midway, findings relayed from CSAW's outpost in Hawaii enabled U.S. naval forces to avoid a trap prepared by Japan and to spring a trap of their own that sank four Japanese aircraft carriers. These back-to-back victories reversed the tide in the Pacific war, throwing Japan on the defensive.[13]

CSAW and all other wartime code breakers worked in the long shadow of the Government Code and Cypher School at Bletchley Park. At its peak the Bletchley Park

FIGURE 1.5. Rebuilt Colossus code-breaking computer (vintage 1944; rebuilt 2007). The British-built Colossus cracked German codes at Bletchley Park. Colossus was entirely secret until the 1970s. Source: MaltaGC under GNU Free Documentation License, version 1.2.

staff, located on a 580-acre manor estate halfway between Oxford and Cambridge and conveniently close to London, formed a small city of twelve thousand. There, cryptography experts designed and built special-purpose computers to automate the cracking of the two principal German codes. The supersecret Colossus machines grappled with the Lorenz teleprinter cipher, which scrambled the five data bits that coded each letter of an encrypted message.[14] The British also built electromechanical sorting and matching machines, known as "bombes," to crack the more difficult Enigma code, with its three or more interlocking rotors that directly scrambled the letters of a message. Alan Turing gained wartime fame for "banburismus," a clever method of logic and guesswork that he invented to winnow down the immense number of possible Enigma rotor configurations to a smaller number that could be tested by the bombes. To take a simple example, Enigma's continually shifting rotors meant that a string "AAA" in a row might be coded "RVQ," but no letter was ever coded directly as itself: "A" might be "Z" but it was never "A." This insight alone ruled out nearly two thousand possible combinations. Additional tricks used by Turing and the other code breakers relied on abstract mathematics to identify recurring "loops" and other subtle patterns. At least three separate stories trace the name of "bombe" variously to the ticking sounds they made while working or an obscure nickname of "bomba" for an early version of the machine. Perhaps it was merely an ice-cream treat enjoyed by the Polish cryptographers who worked out an early scheme to crack the Enigma messages. However the name arrived, it was the Poles' insights, suitably refined and cleverly mechanized, that made possible the electromechanical bombes.[15]

Across the Atlantic, CSAW independently built an American version of the British bombes. Early in the war, the National Cash Register Company in Dayton, Ohio, turned over its Building 26 to the newly formed Naval Computing Machine Laboratory. For a time CSAW's Howard Engstrom took personal charge of its work, then turned over command to Ralph Meader, who was described by an insider as a "super salesman . . . the one who administers things from the standpoint of getting the customer's acceptance . . . [a] public relations kind of a person."[16] For the American attack on Enigma, researchers at MIT with grandiose visions initially proposed an all-electronic machine that called for an unheard-of twenty thousand vacuum tubes (even the army's singular ENIAC computer did not use this many tubes). In Dayton, Joseph Desch, head of NCR's research laboratory, had experience with superfast counting circuits that, at a million cycles a second, could accurately measure the speed of an artillery shell shot from a cannon. He set aside the extravagant MIT design and, working at a feverish pace, led a crack team of engineers from NCR, IBM, and MIT that combined electromechanical rotors with electronic circuits. The American engineers never saw blueprints for the British bombe, as best as anyone can tell, but they did know in a general way what it did. "It seems a pity for them to go out of their way to build a machine to do all this stopping if it is not necessary," commented Alan Turing after a wartime visit to Dayton. "I am now converted to the extent of thinking that starting from scratch on the design of a Bombe, this [American] method is about as good as our own."[17]

In top-secret conditions in Dayton, six hundred women—members of the navy's Women Accepted for Volunteer Emergency Service, or WAVES—built a total of 121 code-cracking bombes. The women worked in three shifts to solder together components for the sixty-four wheels that mimicked the multiple rotors of an Enigma machine. "To maintain secrecy, one WAVE was given the wiring diagram for one side of the wheel. Another WAVE soldered the other side." Even decades later, people who worked in Dayton were unusually circumspect. "Well, there is not too much I'm at liberty to say about it, except that we did work on a Navy contract," noted Desch's

FIGURE 1.6. Member of WAVES with American "bombe" (circa 1944–45). Designed and built in Dayton, Ohio, the electromechanical "bombes" cracked German and Japanese codes at CSAW in Washington, D.C. Source: U.S. Navy.

assistant. "It was a very hush-hush, secret device."[18] As they were completed, the bombes were shipped to Washington, where a legion of women put them through the paces at CSAW's Nebraska Avenue complex. The WAVES transferred from Dayton to Washington saw the complete bombes for the first time only when they began operating them, with their rotors racketing along at two thousand rpm.[19] "When this machine started running, it was so loud . . . all of us got to the point that we wanted to scream that it hurt so badly," one recalled. "We soon learned we had to tune the noise factor out—mentally tune it out. It took us about six weeks because we would sleep and still hear that awful noise."[20]

Taking a leaf from Turing's playbook of code-breaking tricks, CSAW also "played those games with them to cut down our trials [of Enigma rotors]. So we never had to try all the possibilities." When a bombe found a promising "hit," lights blinked, a bell rang, and the bombe's printer spat out the result. "We'd take the sheet of paper down the hall and knock on a door," recalled one of the on-duty WAVES. "A hand would come out; we'd turn over the printout and go back and start all over again." Hits were tested and soon they began yielding real results. One CSAW staffer recalled, with evident pride, "when you did something that used a little ingenuity and got out something [i.e., actionable intelligence] in time to get a submarine sunk just before it came out into the open Atlantic, which I did on one occasion, and it was a submarine tanker . . . to watch the Germans for three months later trying to arrange rendezvous with this tanker we knew had been sunk." CSAW used the bombes, once they had completed their daily diet of Enigma traffic, to attack the difficult Japanese naval codes as well.[21] The experience of running a large enterprise during the war led Engstrom, Norris, Meader, and others to re-create their group after the war as the Engineering Research Associates, the first Minnesota computing venture, as described below.

Shell-shocked

During the war years, everybody who was stuck with a supremely difficult mathematical problem eventually found their way to John von Neumann. "Johnny," to his friends, had grown up a mathematical prodigy in prewar Budapest, a high school classmate of famed physicist Eugene Wigner. Along with Albert Einstein and Kurt Gödel, he was a founding member of the Institute for Advanced Study in Princeton, where, impressed by a young British mathematician's wholly novel approach to a tough problem in number theory, he had mentored Alan Turing on a fellowship. Von Neumann published 150 technical papers in pure and applied mathematics, with major contributions to quantum mechanics, statistics, set theory, game theory, and computer science. His brain worked at lightning speed. The story is told that someone posed to him an "infinite series" mathematics problem requiring extensive computation. Take this example: two slow freight trains, one hundred miles apart, steam toward each other at ten miles per hour, while a speedy bumblebee zips ahead at twenty miles per hour, instantly reversing its course the moment it touches the far locomotive, and repeatedly flying back and forth as the trains get closer and closer and closer. Von Neumann was asked, in one version of the story set at a glitzy Washington cocktail party, "How far does the bumblebee fly?" A mathematician might start work summing together the series $1 + 1/3 + 1/9 + 1/27 + 1/81 \ldots$ and multiply the results by $66\frac{2}{3}$ miles. (A shortcut notes that the bee will continue to fly back and forth at twenty mph for five hours, until the two trains collide, for one hundred miles total.) Reportedly, he thought a moment, then affirmed that, indeed, the sum of the infinite series indicates one hundred miles.[22]

During the war years, von Neumann's travels by train took him a good bit farther than the hapless bumblebee. Among other weighty responsibilities, he was consultant to the top-secret Los Alamos atomic bomb laboratory in New Mexico, where the country's top physicists were wrestling with an impossibly difficult problem in spherical mathematics. The Manhattan Project's plan for a uranium bomb called for a simple gun-type design where one subcritical mass of enriched uranium was fired into a second mass, starting an atomic explosion. Plutonium, however, was unstable in any sizable mass, and the most promising design began with a large hollow sphere. The problem was how to design explosive charges to quickly squeeze the hollow sphere of plutonium into a single critical mass of plutonium. Hoping to locate sufficient computing muscle, he went up to Harvard University to consult with Howard Aiken and Grace Hopper, who were then putting through its paces a giant fifty-foot-long programmable calculating machine. Certainly quicker than a human in adding or subtracting its standard twenty-three decimal-digit numbers (it could do three each second), the Harvard Mark I still required ten seconds for a multiplication and ninety seconds to compute one logarithm. Von Neumann saw that it would not do for designing the spherical bomb. Another, apparently simpler, problem to compute the trajectories of artillery shells flying twenty miles through the air to their target took

FIGURE 1.7. Bush Differential Analyzer at the University of Pennsylvania's Moore School, circa 1942–45. Kay McNulty, Alyse Snyder, Sis Stump, and others computed trajectories for artillery shells, work that led to the pioneering ENIAC computer. Source: U.S. Army.

him to the army's Ballistics Research Laboratory and so connected von Neumann's top-secret world of military mathematics to the Philadelphia story. The connection profoundly changed the path of digital computing.

Oddly enough, the impressive computational facility built up by the University of Pennsylvania remained unnoticed by John von Neumann and even Vannevar Bush, the nation's wartime research and development czar, for many months. Penn's Moore School of Electrical Engineering was home to a room-sized electromechanical computing engine built to a design that Bush himself had developed at MIT. These "differential analyzers" were capable of solving the fiendishly difficult mathematics needed in region-spanning electrical grids, intricate vacuum tubes, even the esoteric physics of cosmic rays. A calculus ace might make some lucky guesses, or even exactly solve systems of differential equations with simple linear terms, even with three or four sets of equations that described the dynamic forces at play. But when two physicists needed twenty-seven integral equations to represent the interaction of cosmic rays with the earth's magnetic field, additional computational muscle was definitely needed.[23] If you could afford it, you might put to work a small army of human "computers" armed with Monroe or Marchand calculators to laboriously compute an approximation. Or, if you had deep pockets, you might rent time on a Bush-style differential analyzer. There were three or four of them in the United States.

Human and machine computers labored at the U.S. Army's Ballistics Research Laboratory, an hour or so train ride south from Philadelphia at Aberdeen, Maryland. It was army policy to send a new artillery gun into the field only after preparing a precisely computed firing table to accompany it. Obviously, a gunner needed to accurately aim the weapon, making appropriate allowances for distance, air temperature, wind speed, even the subtle effects of gravity as the shell lofted high over the earth. The army commandeered the Moore School's differential analyzer and set it to work computing shell trajectories. In addition, two hundred human computers—the recruiting poster solicited "Women with Degrees in Mathematics"—used state-of-the-art Marchand hand calculators to meticulously compute data for the all-important firing tables. It was heady, real-world mathematics. "As the bullet travels through the air, it is constantly being pressed down by gravity. It is also being acted upon by air pressure, even by the temperature. As the bullet . . . got down to . . . the speed of sound, then it wobbled terribly. . . . So instead of computing now at a tenth of a second, you might have broken this down to one-hundredth of a second to very carefully calculate this path as it went through there . . . when you finished the whole calculation, you interpolated the values to find out what was the very highest point and where it hit the ground."[24] One of the supervisors overseeing the women's work in a Moore School classroom, Mary Mauchly, was married to an ambitious physics professor named John Mauchly. You might even imagine that, over dinner some evening, the two of them hatched the first glimmers of what became ENIAC.

John Mauchly, a physics Ph.D. from Johns Hopkins University, had been doing calculations on sunspots and weather data for several years before coming to the Moore School in 1941 for a wartime course in electronics. Teaching undergraduates at a small college outside Philadelphia could not match the excitement of being at one of the country's top technical universities, and when his class finished up he readily accepted an offer to stay on at the Moore School. "Mauchly all the time, even as a student, and for the two years from then until the ENIAC work started[,] was a pest talking about . . . the computer he wanted," recalled one colleague.[25] While Mary went to work supervising the women calculators, Mauchly and a newly minted electrical engineering graduate, one J. Presper Eckert, began scheming about how to build an electronic computer using vacuum tubes. It seemed natural to the two men to adapt the "ring counters" that physicists had been employing to automatically count cosmic ray hits or radioactive decays. Such a ring counter could be fashioned into a superfast adding unit, which they termed an "accumulator," that could directly do the work of addition and, with some modifications, handle as well subtraction, multiplication,

FIGURE 1.8. ENIAC and successor Army computers, 1946–62. From left: Patsy Simmers, holding ENIAC board (1946); Gail Taylor, holding EDVAC board (1949); Milly Beck, holding ORDVAC board (1951); Norma Stec, holding BRLESC board (1962). Source: U.S. Army.

and division. A special control circuit even allowed the machine to take square roots. It summed up the first n odd numbers, neatly computing the square of n, and then worked backwards to the square root. Such a calculating machine offered a promising way to automate the army's ballistics computations, and with an army contract in hand by spring 1943 they were at work on an Electronic Numerical Integrator and Computer; for short, this was ENIAC.

The decimal-based counting scheme evidently was so compelling that the ENIAC team stuck with it even after Mauchly made a four-day visit to a physics colleague at Iowa State, who was working on a prototype binary-based computer (see chapter 5). Eckert and Mauchly's design for ENIAC needed a bank of thirty-six vacuum tubes to store each single decimal number, ten such banks were required to build up one "accumulator" with ten decimal digits, and ENIAC had a total of twenty accumulators that could be wired together for computations. Circuits for timing, control, and card reading required additional tubes. ENIAC's eighteen thousand vacuum tubes, working at electronic speed, could complete 350 multiplications in one second. "A skilled person with a desk calculator could compute a 60 second trajectory in about 20 hours; the Bush differential analyzer produced the same result in 15 minutes; but the ENIAC required only 30 seconds, less than the flight time," the army boasted.[26] It's an apocryphal story that the lights dimmed in West Philadelphia when ENIAC was first switched on, but it was true that for their postwar computing efforts Eckert and Mauchly searched out factory buildings in Philadelphia that offered industrial-grade electricity supply.

When John von Neumann paid his first visit to ENIAC in the fall of 1944, the great mathematician's first question was on everyone's mind. By that time, Eckert and Mauchly were at work on a second-generation "stored program" computer, quite different from the ENIAC calculating machine. Indeed, a month or so before his visit to ENIAC, von Neumann had sat in on the advisory board that blessed the successor project. This was EDVAC, a much more influential creation on paper, without question, than the troubled physical machine ever turned out to be in reality. A stored program computer would not only keep data in electronic storage but would also place its instructions there; the design transformed special-purpose calculating engines into general-purpose machines that could play chess or design bombs. True to form, when he arrived, his first question to ENIAC's designers was about its logical design. Von Neumann's signature concepts of "memory," "control unit," and "arithmetic-logic unit" were, so far as anyone can tell, jointly developed in collaboration with the ENIAC team; but when it came to publish, the credit flowed solely to the great mathematician. Herman Goldstine, who had launched ENIAC by lining up army funding, assembled a set of von Neumann's letters, put von Neumann's name on the cover of the 101-page report, and on June 30, 1945, with "First Draft of a Report on the EDVAC," created the founding document of modern computing.[27]

FIGURE 1.9. Programming ENIAC at the Moore School. The two women working on ENIAC were cropped out for a 1946 army recruiting advertisement, in which only Corporal Irwin Goldstine (foreground) remained after the man in back was also removed. Source: U.S. Army.

For its part, ENIAC was blindingly fast at addition but it was not "programmable" in the present sense of the term. The job of getting ENIAC to do useful work fell to six women recruited from the ranks of the army's human computers. Eckert and Mauchly gave them a stack of circuit diagrams, staff members told them how accumulators should work, and much of the rest was their own invention. As Betty Jean Jennings recalled, "The biggest advantage of learning the ENIAC from the diagrams was that we began to understand what it could and what it could not do. As a result we could diagnose troubles almost down to the individual vacuum tube. Since we knew both the application and the machine, we learned to diagnose troubles as well as, if not better than, the engineer."[28] The women used plug boards and patch cords to connect accumulators in the correct patterns to solve the mathematical equation that was put to the machine. With the war winding down, ENIAC's first significant problem was done for Los Alamos to investigate a thermonuclear hydrogen bomb.

FIGURE 1.10. Military officials and men of the ENIAC team, 1946. From left: J. Presper Eckert Jr., chief engineer; J. G. Brainerd, supervisor; Sam Feltman, chief engineer for ballistics, Ordnance Department; Captain H. H. Goldstine, liaison officer; John W. Mauchly, consulting engineer; Dean Harold Pender, Moore School of Electrical Engineering; General G. M. Barnes, chief of the Ordnance Research and Development Service; Colonel Paul N. Gillon, Army Ordnance Research and Development Service. Source: U.S. Army.

At its public unveiling in February 1946, ENIAC was surrounded by the military officers who had funded it, and the men who had designed it, but not the women who made it do useful work. As recalled by one of the leading participants, Herman Goldstine, in his book *The Computer from Pascal to von Neumann* (1972), "The actual preparation of the problems put on at the demonstration was done by Adele Goldstine and me with some help on the simpler problems from John Holberton and his girls."[29] Only fifty years later, in 1997, were these women—Kathleen McNulty Mauchly Antonelli, Jean Jennings Bartik, Frances Snyder Holberton, Marlyn Wescoff Meltzer, Frances Bilas Spence, and Ruth Lichterman Teitelbaum—given proper recognition for their intellectual achievements and wartime efforts.

Five months after ENIAC's public demonstration, nearly everyone in the nascent field of digital computing attended the famous Moore School lectures. For eight weeks in July and August 1946, members of the Moore School team presented forty-eight morning lectures and afternoon demonstrations before a select audience. Eckert and Mauchly, having left the university a few months earlier, each received $1,200 for their teaching, while others received travel expenses and fifty dollars per lecture. Among the twenty-eight official attendees, the navy and MIT sent the largest delegations (six each), followed by the National Bureau of Standards (three) and other military agencies; General Electric, Bell Telephone Laboratories, and Reeves Instrument Company were among the government contractors. A unique set of detailed notes by MIT's Frank Verzuh allows us to vicariously sit in the audience. The Office of Naval Research published most of the forty-eight lectures, although von Neumann's on August 13 was never published; some accounts, erroneously, state that it was never given. Verzuh's notes are all we have on the great mathematician's treatment of "New Problems and Approaches." The topics he favorably discussed—and some he tried to discourage—are of interest in understanding the new field of computing and how secrecy shaped its growth.

In his Moore School lecture, von Neumann outlined the contours of the emerging field. Computers, he suggested, could directly deal with wind-tunnel simulations, systems of linear equations, and three-dimensional meteorology. When an audience member questioned his optimism also on partial differential equations for hydraulic turbulence, von Neumann reaffirmed that he was "confident that it could be done." After all, these were problems that he was already working on. But, curiously enough, with a sharp warning about the difficulty of combinatorial problems, he pointedly deflected attention away from the applications of computing for code breaking. (Ironically, at least one actual code breaker in the audience took away precisely the opposite lesson, penning a secret report that led to Minnesota's first computing machine, as chapter 2 details.) By mid-August, when the excitement about EDVAC tailed off, the demanding students, with "boisterous denouncements of the manner in which the course was being conducted," repeatedly demanded and finally received a week-long demonstration of ENIAC.[30] After all, it was at the time the world's largest working electronic computer.

One hesitates to suggest that Eckert and Mauchly's commercial venture to build a commercial successor to ENIAC was doomed from the start. A dispute over patent rights had forced them to leave the Moore School. They founded a new company in Philadelphia to build a product that did not yet exist in an industry that no one had heard of. Philadelphia had long specialized in the textile, chemical, and machine-shop industries, but it was not so advanced in the electromechanical industries such as electric locomotive or streetcar building. While ENIAC had cost roughly five hundred thousand dollars to build, they began taking commercial orders for an even more complicated machine—a full-blown stored program machine they would call UNIVAC—for pitifully small sums. Supremely naive about how to run a business, or just possibly desperate to get an additional contract in hand, they stood to lose two hundred thousand dollars or more on each and every order. Even though no one at the time knew what a computer might cost, Mauchly and Eckert were somewhat reckless with their financial calculations. For a later computer, as one veteran recalled, "Pres Eckert asked each of his engineering managers to come up with quotes on their part. . . . And he looked at the total of those figures and said, 'That's totally unrealistic. We could never sell that. Go back and reduce your estimates by half.' When we said, 'This is unrealistic,' he said, 'Well, it doesn't matter. Once they get the taste of it, we can then talk about a new price.'"[31]

As it turned out, most of Eckert and Mauchly's ruinous contracts were eventually set aside as unworkable, and so accordingly they had all the expenses of developing a new computer without any of the revenues that might result from successfully delivering one. In addition, they got behind in their paperwork and did not promptly file for a patent on ENIAC. The concept of a computer, according a later lawsuit (examined in chapter 5), was publicly disclosed with von Neumann's "First Draft of a Report on the EDVAC" in June 1945. Then the unveiling of ENIAC itself came with a press conference and demonstration in February 1946. Patent law at the time required them to do an initial filing not more than twelve months after the first public disclosure—certainly no later than February 1947—but in fact they waited to file for a patent until June 1947. And if you are wondering where the one hundred thousand-dollar grant from the army to build ENIAC's stored-program successor ended up, it stayed at the Moore School.

Eckert and Mauchly struggled from nearly the moment they left the Moore School until February 1950, when, essentially bankrupt, their company was bought up by Remington Rand. Simply put, the costs of building a pioneering commercial computer were vast, and they had inadequate income and insufficient capital. A 1948 contract with the Census Bureau was promising; and the first UNIVAC was, in March 1951, in fact "delivered" to the Census Bureau even though it stayed put in Philadelphia for almost two years while crunching out census data. Eckert and Mauchly quickly signed up when Northrop Aircraft sought an "experimental computer . . . to prove the feasibility" of a guidance system for the air force's long-range Snark missile. What might best be considered as a one hundred thousand-dollar development contract resulted in their delivering BINAC, a 1,400-tube prototype that in spring 1949 was quite possibly the first stored program computer to successfully operate in the United States, but that machine was crated up and shipped off to Northrop in California. After that, nothing positive was ever heard of BINAC, which arrived in Hawthorne, California, in "deplorable condition" (according to Northrop) and suffering from "unreliability and insensitivity" among two dozen serious deficiencies. Even

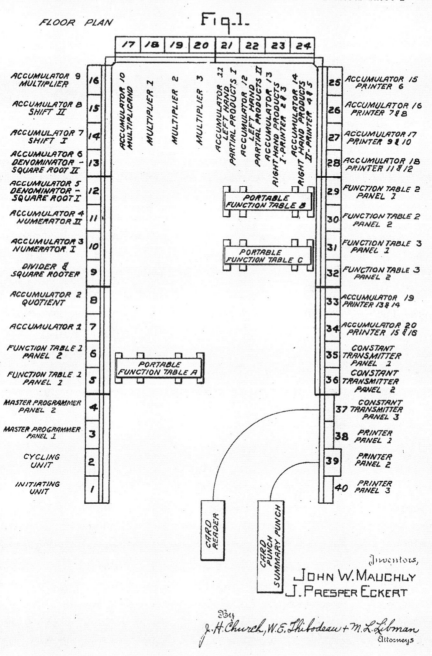

FIGURE 1.11. Eckert and Mauchly's patent on ENIAC, filed June 26, 1947, was overturned in *Honeywell v. Sperry Rand* (1971–73); see chapter 5. This top–down view shows ENIAC's twenty accumulators and control units. Source: USPTO.

on this ill-fated effort Eckert and Mauchly's expenses ran to $270,000, and the resulting red ink compounded their financial plight.[32] Sometime during these troubled months, Eckert and Mauchly delivered a product "to the Army Security Agency . . . under a special contract we had with them for some classified equipment," but details about it are not publicly known.[33]

Eckert and Mauchly gained a year or so of breathing room when a racetrack betting company ponied up five hundred thousand dollars in financing, but that respite abruptly ended when a plane crash in October 1949 killed their benefactor in the company and its investment was called in. To make matters worse, word somehow got around that Eckert and Mauchly employed politically suspicious engineers. Perhaps Mauchly himself had the wrong friends. The McCarthy-era plague settled on the company, stripping Mauchly of his government security clearance during a critical period and quashing at least two potential sales prospects with the navy and the nuclear facility at Oak Ridge, Tennessee. Mauchly's security troubles began in April 1948 and caused problems even after the Remington Rand acquisition in 1950, because, according to historian Nancy Stern, "the old security problem that had plagued Mauchly several years before had never been fully resolved."[34] It took years for Mauchly to eventually clear his name; in the interim the firm's finances ran aground. The takeover offer arriving from Remington Rand, while not especially generous, must have been a relief. Eckert and Mauchly now had jobs that paid regular salaries.[35] Their debts were paid. Within two years they would be joined in the Remington Rand collection of companies by a second group of would-be computer engineer-entrepreneurs who, despite their own best efforts, found themselves in a similar scrape.

CSAW to ERA

Even while the fighting wore on in the Pacific and European theaters, and CSAW continued its top-secret cryptography activities, Engstrom and Norris began to consider the possibilities of what might be ahead. They had created a high-spirited group of unusually talented technical experts, and they hoped somehow (as a colleague put it) "to preserve the tremendous resource that was assembled during the war in the intelligence community."[36] Although they couldn't know it at the time, they were looking to create something like the RAND Corporation, the great research and development enterprise originally formed in 1946 under the protective wing of the air force to develop such forward-thinking notions as "Preliminary Design of an Experimental World-Circling Spaceship." The idea of doing technological research—independently from a university campus or even a government laboratory like the National Bureau of Standards—was novel and untried. At its origin RAND depended on well-placed friends in the air force as well as Douglas Aircraft, an essential intermediary to handle the intricacies of government paperwork. Minnesota's history in computing began when Engstrom and Norris mobilized their contacts in the navy and happened on a fortuitous contact that led them to an abandoned glider factory in St. Paul.

The war years gave Engstrom and Norris the heady experience of running a large and successful operation. The last thing either of them wanted was to go back in time to teaching mathematics at Yale or selling Westinghouse X-ray machines. Their wartime work, while not exactly in the public eye, had gained the attention of some of the movers and shakers in Washington. Joseph Wenger, for one, the commanding officer of CSAW and Engstrom's boss, had successfully centralized naval intelligence and was keen on preventing his effort from being scattered to the wind. After the war, he climbed into the executive ranks of military intelligence, serving as number two for the National Security Agency at its founding in 1952. He was well placed to steer a contract or two their way—and, to Minnesota's everlasting benefit, did so.[37] James Forrestal, for another, a Wall Street wunderkind before the war and then secretary of the navy during the war, also saw the merits in sustaining CSAW. For a man who had snatched the first flag posted on Iwo Jima for his own souvenir (the famous photo that everyone remembers was taken of an oversize replacement flag), and was soon to be named secretary of defense, Forrestal made a disappointingly tame suggestion to Engstrom and Norris of civil-service appointments.

Engstrom circulated a secret seven-point plan in February 1945 that contained the seed that became St. Paul's ERA. The plan proposed an entirely new National Electronic Laboratories to do engineering development for naval communication and intelligence, pointing to the precedent of successful wartime cooperation between NCML in Dayton and CSAW in Washington. The new NEL would corral the "best collection of talent in the world," believed to be superior to the British as well as the army efforts, to achieve the "essential" goal of carrying on research and development for navy intelligence. Its location was, for the moment, tantalizingly vague. Ongoing top-level cryptography work at NCML, Bell Laboratories, and Eastman Kodak would be folded into the company, to be headed by four prominent CSAW and NCML staffers—including Engstrom himself. Despite the assertion that "private funds are available to provide the necessary facilities," actual prospective investors told Engstrom and his group to forget the foolhardy idea of founding a brand-new firm to do top-secret R&D.[38]

Although everyone appreciated wartime code breaking, it seemed that no existing company wanted to continue this specialized activity into peacetime. IBM, National Cash Register, and Eastman Kodak each had devoted significant resources to cryptography during the war effort. "If one or two of these big companies had really cozied up [to government contracting], there never really would have been a need for ERA. But, they all had their fish to fry and they didn't make that much money, I guess, on these code-breaking machines," thought James Pendergrass.[39] In Dayton, the shift to peacetime production was especially pronounced. Despite Desch's brilliant engineering on the American-style bombes, and the successful wartime production of more than one hundred of the special machines, NCR wanted nothing to do with postwar cryptography or military electronics. As Desch's right-hand man Robert Mumma recalled, "The management at NCR was not very happy about all this war work because it was taking so much of their effort. They wanted to get back into their cash register business, their accounting machine business, and that sort of thing. So . . . [after the war] they didn't agree to any outside contracts."[40]

At this point, with prospects for a new company rapidly dimming and peacetime demobilization at full clip, John Parker providentially appeared. Parker was another investment banker, a board member of Northwest Airlines, active before the war in financing several aviation companies. He had inherited the Washington-area offices of a New York Stock Exchange brokerage and, through his well-placed Annapolis classmates and well-heeled investment buddies, was rather well connected in wartime

FIGURE 1.12. John Parker, founding president of Engineering Research Associates. The photograph on the wall shows one of the large wooden gliders manufactured at Parker's Northwest Aeronautical Corporation in St. Paul, later the ERA factory.

Washington. "Parker was . . . one of the greatest salesmen I ever met, and he was positive and charming when he wanted to be. When he wanted to do something, he had a great ability to charm the socks off of whoever was involved," recalled one colleague.[41] When in St. Paul, he and his wife lived in the fashionable Commodore Hotel. Richard Lilly, chairman of the First National Bank of St. Paul, knew that Parker was someone who could get things done. He tapped Parker to rebuild and revamp the Toro Manufacturing Corporation of Minnesota in 1945. "The old Toro plant had one long rod that ran from one end of the place to the other, and every machine was hooked up on this one rod. Well, we went in and modernized the plant and put modern equipment in and so forth," Parker noted.[42] Owing something to Parker's organization and savvy, Toro became a national force in manufacturing and marketing lawn mowers and snowblowers.

But Parker, too, had a pressing problem at the time. During the war years he had a tidy business in manufacturing wooden gliders in St. Paul. Wooden gliders were ideal for silently evading enemy spotters and fourteen thousand of them were used in Europe and Asia for reconnaissance and even one-way missions dropping troops and equipment into battle. Even though his Northwest Aeronautical Corporation built 1,510 copies of the most popular design, the second-largest national output behind only the Ford Motor Company, Parker knew that this profitable run had ended. With peacetime at hand, no one needed the oversize boxy wooden gliders. His problem was finding something else for the St. Paul factory, and so he listened attentively when the navy men knocked on his door with a promising, if vague, suggestion. He soon met with Fleet Admiral Chester Nimitz, chief of naval operations, who poked him on the chest and declared, "there's a job that I would like to have you do . . . It may be more important in peace time than it is in war time." At some point, a copy of the Engstrom memo came into his hands, with the classification "secret" neatly blocked out. "It was like taking on the Symphony Orchestra without knowing a note," Parker recalled, but he agreed to organize the new venture.[43]

The invention of the "Engineering Research Associates" in January 1946 was the equivalent of the skeleton for a skyscraper, with such small details as a foundation and walls to be added in due time. The skeleton consisted of the modest sum of twenty thousand dollars paid in capital, half from a group of investors that John Parker had brought together and the other half from Howard Engstrom, Bill Norris, and an additional forty-seven "associates" from CSAW and NCML who invested at ten cents a share. The agreement in effect bound the stockholding associates to ERA for two years. In addition, Parker arranged for a $175,000 line of credit for working capital. It was illegal for any such new entity to receive government contracts of any significant size, and so the Northwest Aeronautical Corporation continued, at least on paper (until it was wound up in 1948), to be the official contracting entity.[44] Wenger helpfully sent two early contracts to ERA to get things going. It is not

FIGURE 1.13. Engineering Research Associates, circa 1950. John Parker's wartime Northwest Aeronautical Corporation glider factory became the first plant of ERA, surrounded by fencing because of its classified military work.

recorded how the forty CSAW staff who were transferred, accustomed to being at the center of things in wartime Washington, responded to their new location in an industrial district of St. Paul.[45] The main building, built as a radiator foundry in the 1920s, had open windows and so "sparrows and swallows would get in and fly around the high ceiling. It was necessary periodically to chase birds out of the building. In the winter, the inside temperature dropped so low . . . that programmers wore overcoats and mittens at their desks."[46] Another denizen remembered "that god damn old plant full of sparrows that crap on everything."[47] And just in case anyone was uncertain about the navy's imprint—with a navy-affiliated cryptography group and the Naval Computing Machinery Laboratory transferred from Dayton, Ohio, in summer of 1946—the St. Paul factory was leased from the navy and officially designated as a Naval Reserve base so that the navy might post guards there to keep watch over the goings-on. With the Engineering Research Associates, a highly promising enterprise had landed in Minnesota, but it was far from clear whether ERA would have the resources to take off in the new world of electronics and computing.

2
St. Paul Start-up

Engineering Research Associates
Builds a Pioneering Computer

The place in St. Paul where Minnesota's computing industry was born neatly connected the state's prairie history with its high-technology future. When you go there, even today, you can easily imagine the distant echo of lumber, cows, and horses—and you can readily hear the railroads, pounding out their rhythms of commerce and industry. The state's pioneering computer factory was a half mile from the Midway industrial district's epicenter at Snelling and University, sandwiched between the tracks of the Great Northern and the Chicago, Milwaukee & St. Paul railways. The surrounding blocks had only recently emerged from turn-of-the century wooden sidewalks, wire-fenced farms, and "prairie land" that pastured cattle for two local dairies. Brooks Brothers Lumber Storage had "green lumber shipped in [and] piled to dry," while directly opposite the future computer factory on Minnehaha Avenue had been offices, a church, and a "cast iron horse drinking fountain." Cows drank at a nearby fountain across Snelling near the present location of Hamline University. The St. Paul Union Stockyards, founded in 1886 as a temporary holding site for cattle destined for the South St. Paul slaughterhouses, constructed two years later by the Chicago railroad magnate A. B. Stickney, began at the parcel's west boundary along Prior Avenue. At the south boundary along University had been a stockade where "Barrett Zimmerman broke wild horses."[1]

In the 1920s, Midway was developed as a manufacturing and industrial district, while retaining something of its mixed commercial and residential character. The Union Stockyards became a switching yard that shuttled freight cars between no fewer than nine major railroads. "When the four railroads to the Pacific coast were completed, all freight from the West was routed through what was called the Minnesota Transfer in the Midway district. Around this transfer developed an industrial and commercial center and blocks of residences for the workers. Through it were carried the goods of the Orient, the lumber and fruit from the Pacific States," said the *WPA Guide to Minnesota*. Before the Panama Canal opened in 1915, the Midway district "was the scene of extraordinary railroad and shipping activity."[2]

The Midway district created a dynamic urban cluster of people and skills as well as freight. The Twin City Rapid Transit Company established offices and a major repair and maintenance facility immediately south of Snelling and University. Montgomery Ward purchased twenty acres from the transit company to build its Midway store and distribution center, adding to the district's vitality. Over the years, the transit company's Snelling Shops complex—a major industrial installation taking up most of the forty-acre property with sizable forge, foundry, powerhouse, carpentry, erecting, and machine-shop capabilities, in addition to the repair facilities—made more than a thousand streetcars to a specially adapted design for the region's hard winters, shipping them to half a dozen western cities. The Snelling Shops built up an important urban locus of electrical and mechanical skills and expertise.[3] Minnesota's early computing industry flourished owing to its location in a skill-rich industrial district.

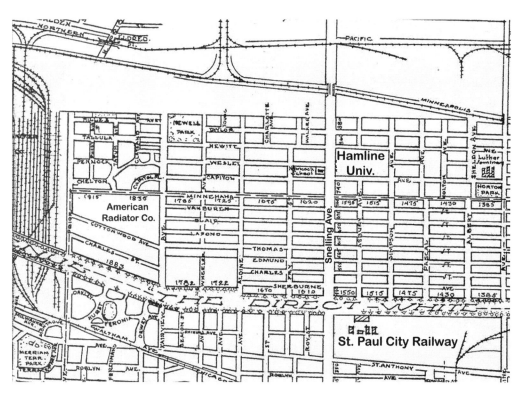

FIGURE 2.1. St. Paul's Midway industrial district, 1923. The ERA plant on West Minnehaha Avenue was located next to the rail yard of the Minnesota Transfer Railroad Company, formerly the Union Stockyards (far left). University Avenue is labeled as "The Direct White Way." Source: Hamline–Midway History Corps.

The district's industrial vitality generated notable commercial activities and even some architectural distinction. The Griggs–Midway building at 1821 University was a canning factory "said to be the largest of its kind in the world" that was further expanded in the 1920s into a "huge concrete frame building."[4] The pioneering computer factory nearby was originally built in 1924 for the American Radiator Company as a foundry, assembly site, and warehouse. When the radiator company closed up six years later, the Civilian Conservation Corps moved in a regional quartermaster depot. Other companies that made candy, crackers, paper products, and adhesives also built or expanded factories in the district, while banks and commercial establishments lined up along Snelling or University Avenue, which, after it was ostentatiously lighted up in the mid-1920s, became known as the Direct White Way with trolley cars linking the Twin Cities. In the 1950s, University Avenue, after the trolley cars were retired, became a prominent "auto row" with a dozen automobile dealerships and sixty trucking firms. The industrial district was undergoing a significant transition, a harbinger of the region's technology base and future economy. Machinists who had once used precision lathes to profile trolley-car wheels would soon be machining large cylinders for magnetic-drum memory units: they would employ tiny magnetic pulses to store digital information. In the Midway industrial district, Minnesota took its first steps toward becoming a digital state. The intelligence-and-electronics complex described in chapter 1 took new form, in the early postwar years, as the Engineering Research Associates, known then and now as ERA.

The transformation of Midway into a high-tech industrial district began within a few months of ERA's founding in early 1946, when cryptographers from CSAW in Washington, officers from the Naval Computing Machine Laboratory in Dayton, and a dozen or so recently minted engineering graduates began assembling at 1902 West Minnehaha Avenue. Before long, ERA hired fully 40 percent of the University of Minnesota's electrical engineering class of 1943, including Frank Mullaney and Erwin Tomash, and a smaller but notable fraction of the postwar graduating classes, including Seymour Cray, Jay Kershaw, and Jim Thornton. "I got through security stuff, they had high clearance requirements which were in place, they had areas where I could work until all of that was done. My first desk was up in a balcony in just a terrible looking place with a chair with three casters, an old beat up chair to sit in and so on and not much else," one new recruit recalled. Security officers initially handed out identification badges from Northwestern Aeronautical (NAC) "with the little wings on them . . . until we got ERA credentials."[5]

Navy security officers vetted new arrivals and all visitors in a squat rectangular building (located in the photograph [Figure 2.2] close to the upper parking lot and along the street). A two-story office building lay between the parking lots. Machine-shop equipment was installed in a long, low-slung building with tall light-friendly windows (that runs diagonally starting at lower left). An assembly hall was at the

48 ST. PAUL START-UP

FIGURE 2.2. Engineering Research Associates plant number 1, circa 1955. Minnehaha Avenue runs east–west along the right side, while Prior Avenue runs north–south at the top of this view. Sperry Univac, ERA's successor, occupied these buildings until 1991.

opposite end (the low square building in upper right). "There were a series of railroad type rooms; you could walk from one to another," recalled one engineer. Secret projects were isolated in building 6, a specially fenced-in single-story building (just below the office building near the lower parking lot). Veterans recall that the really top-secret projects were housed in that building's basement out of sight of all prying eyes. One remembers being subjected to a polygraph exam to gain entry, and, to this day, maintains that the "project that required that special clearance is not open for discussion."[6] A second main building to the complex (in the upper left) was later known as Univac plant 5. ERA, like NAC before it, leased the eleven-acre facility "for $1.00 a year plus maintenance" as Navy Industrial Reserve Plant number 196. It remained a leased navy property for nearly fifty years, until 1991 when the ERA-successor Univac finally closed its lease and left the property. Even today, with the buildings largely abandoned, one still might sniff the lingering scent of metal shavings and electricity in the air. It requires no stretch of the imagination, walking along on a hot afternoon, to remember that, as one inhabitant phrased it, "In the summer, the temperature rose to the point that whole departments worked without shirts on, and draftsmen [had] trouble with sweat dripping on their drawings."[7] Finances for the new concern were touch and go, kept solvent by the good graces of Richard Lilly, president of Northwest Airlines and chairman of the First National Bank of St. Paul. Joe Walsh, the new company's controller, was "exultant when we got to the point where we didn't have to borrow money every month but only every few months."[8]

Cold War Codes

Government contracting in ERA's first years was the lifeblood of the company and a subtle blend of secrecy and legerdemain. "It [ERA] had relatively little market for a product that was sold to other than government customers or military customers in the early stages," noted one engineer. "They would get a contract from the Navy or the National Security Agency or from the Air Force to do a certain thing . . . , almost all . . . cost plus fixed fee type of business."[9] During the war, CSAW had been a top-secret agency that avoided the public eye, and it remained that way even after the war; its Washington-based staff and activities were folded into a transitional joint army–navy intelligence agency that eventually emerged in 1952 as the National Security Agency, with CSAW's former chief Joseph Wenger as the vice (deputy) director. The NSA, nicknamed "No Such Agency," studiously avoided the limelight and it too arranged external contracts through other branches of the military services, such as the navy's Bureau of Ships, even when the contract activities were central to its intelligence mission and had nothing to do with ships. "The intelligence agencies [including] NSA," as one CSAW staffer phrased its habit of convoluted contracting, "they've got some squirrely set-ups."[10]

Other tricks were employed by the ERA staffers in St. Paul and military officers in Washington to avoid the ire of the National Bureau of Standards, which was vigorously and publicly pushing its own agenda that would lead to the early stored program computers SEAC and SWAC. "They were very careful, our sponsors, never to call these things computers," recalled ERA's Arnold Cohen. To ensure that its computing activities stayed beneath the NBS radar, the NSA labeled its early computers "analytical machines," which sounded, safely enough, more like Charles Babbage's nineteenth-century mechanical adding machines than the state-of-the-art electronic digital computers that they actually were developing. Special features might even be added to a machine as "part of a little game that made it classified" and thus clearly beyond the reach of the civilian-only NBS.[11] All of this contractual craftiness had the effect of disguising the public history of Minnesota computing.

Especially after the NCML staff from Dayton joined the CSAW cryptographers, St. Paul became one of the world's leading sites for state-of-the-art cryptography and, accordingly, the early history of digital computing. In Dayton—both before the bombe project came to NCML and after the manufactured bombes were shipped to Washington—a core group of technical specialists worked on several promising code-cracking schemes. Some of the specialists worked along lines suggested by MIT's Vannevar Bush in the 1930s, with his notion of "selecting" and "comparing" pieces of textual data drawn from an immense database, although he did not yet use that particular term of art. Another promising line of development that combined microfilm, electronic, and digital techniques was known as the Navy Rapid Machines program.[12] After the 120 NCML-designed bombes were sent ticking to Washington, the Dayton experts resumed their conceptual cryptographic work. It is no accident that the first several significant projects for which ERA won navy contracts were a mix of cryptography and computing. Some participants' reminiscences even credit NCML for the pioneering Atlas computer, rather than ERA, which speaks volumes to NCML's substantial profile and separate identity.[13]

ERA's initial cryptography and computing contracts had such colorful code names as Alcatraz, Demon, Goldberg, Hecate, and Warlock and more than ample funding owing to the emerging Cold War between the United States and the Soviet Union. Very early, as one ERA veteran recalled, "we soon realized that we were making frequency counts of characters and you soon know what the business is [i.e. code

FIGURE 2.3. Engineers for Engineering Research Associates at work at Minnehaha Avenue plant, 1952.

breaking], . . . there was a big security atmosphere." The Goldberg project, which proceeded in two phases between 1947 and 1951, explored the use of magnetic drums to store, retrieve, and manipulate electronic forms of data. With a magnetic drum measuring thirty-four inches in diameter—sculpted by highly skilled machinists (see photo)—the ERA engineers wrote data onto it stepwise one row at a time, while reading off data while the drum rotated at a snappy fifty rpm. There was a "whole matrix of AND circuits . . . combined in various ways for output. It was a very special purpose kind of thing. Involved an awful lot of tubes and wires," recalled one engineer. "You had to kind of make up things as you went along. There just wasn't a whole lot of information." Recalled another, "We were very impressed with ourselves because the surface was moving along I suppose something like sixty miles per hour, and yet you could very accurately pick out a given bit position and change it on the run. We were very impressed that we could do this very accurately, and reproduce it every time." The latter phase of Goldberg added photoelectric elements for paper-tape scanning and thirty-six decimal-based counting circuits, utilizing a total of seven thousand vacuum tubes.[14] In its tube count Goldberg was nearly half the size of ENIAC.

The short-lived success of one singular code-breaking machine, named Demon, led CSAW–NSA to alter its commitment to all such special-purpose machines. Demon was an even larger eight thousand-tube device that was constructed in eight or nine months as a crash project—forcing "the disruption of all other activity at ERA." "We picked up some 18 people from various other projects and proceeded to work 18 hours a day, 7 days a week," one contributor recalled. "It was just a killer." Demon's circuits created paths to move data from paper tapes onto magnetic drums, and they also shifted data on the drums, searching for suggestive matching patterns. Such matching patterns were the clues to cracking codes when using the brute-force methods of tabulating machines and early computers. While the wartime tabulating machines and bombes had examined specific coding configurations, Demon took a preliminary step toward transforming the cryptography problem into a process of full-blown general-purpose information processing. "There were a lot of elements in [Demon] of things that went into computers like the magnetic drum, tape reader, tape punch, line printer, things of that sort," stated one engineer.[15] With the Demon project, precisely as CSAW had directed, ERA built a machine that was aimed at cracking a specific Soviet code. It was ready for action by October 1948 and it worked brilliantly—for four months—until the Soviets changed their code and made Demon essentially worthless.

A rival line of thinking, inadvertently encouraged by John von Neumann in his Moore School lecture in Philadelphia, suggested a general-purpose computing device that might crack a variety of codes. When the enemy's code changed you would need only to alter the computer's software rather than junk an expensive piece of hardware such as the eight thousand-tube Demon. ERA fully participated in the great postwar

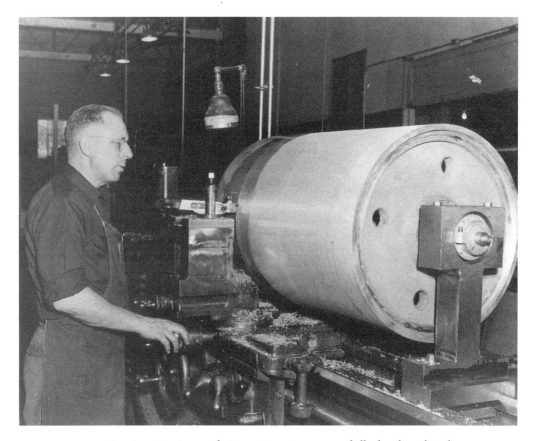

FIGURE 2.4. Engineering Research Associates, circa 1950. Skills developed in the Midway industrial district helped make ERA's magnetic drum memory units.

flurry of enthusiasm for digital computing, with staff members or contacts attending each of the three major conferences—at the Moore School at Penn, Harvard, and MIT—where wartime advances, often enough top secret, were shared among the leading institutions. Curiously, you won't find ERA on the official list of thirty or so Moore School attendees, which came to form a virtual "who's who" in early digital computing.

All the same, ERA was in direct contact with the Moore School through James Pendergrass. During the war, Pendergrass had been one of Engstrom's operational code breakers, and his affiliation on the Moore School attendance roster is the navy's coded term for CSAW. Philadelphia, as Pendergrass himself recalled, was a decisive moment: "It was only when I got to the Moore School Lectures and Mauchly showed us about programming that it became clear to me anyway that what this was [was just] moving bits around. And somehow you made numbers out of these bits. But, really, a computer just handles [and] manipulates bits. We had been manipulating bits at Nebraska Avenue [i.e., CSAW] all during the war. That's where the light dawned."[16]

Von Neumann's strange warning against using computers for combinatorial problems such as code breaking only piqued his interest. After the Moore School lectures, Pendergrass as a staff liaison to the Office of Naval Research visited von Neumann's computer project in Princeton, attended the Harvard computing conference in 1947, and then took a leading role in the conceptual design of ERA's Atlas computer. He worked so closely with the ERA engineers that he is sometimes misidentified as an ERA employee. Oddly, even though his name appears in Wikipedia, his name is blacked out in NSA's declassified history of these events.[17]

Million-Dollar Machine

A breakthrough for ERA was an August 1946 contract from the Office of Naval Research to investigate the "the status of development of computing machine components." At the very moment when people went home from the Moore School lectures in Philadelphia that summer, ONR sent ERA on an extended fact-gathering mission across the country. For a start-up company intent on moving into electronic digital computing, it was a godsend to have a navy contract to ask pointed questions of everyone working in this emerging field. ERA staff made visits to Eckert and Mauchly in Philadelphia, von Neumann's computing project at the Institute for Advanced Study in Princeton, the pioneering SEAC at the National Bureau of Standards, and university sites, including MIT, Brown, Penn, and Harvard. They also reviewed a number of classified governmental reports. "So we knew just about everything that was going on in the USA. We didn't get any thing from the British. Only later on did we find out what they were doing. But nevertheless, I think ERA was privy to practically everything that went on in the US during that year and a half or so."[18] The report to ONR, overseen by ERA's vice president for research, Charles B. Tompkins, and commercially published by McGraw–Hill in 1950 as *High-Speed Computing Devices,* eventually became the hardware "bible" for early computing in much the same manner as Wilkes, Wheeler, and Gill's *The Preparation of Programs for an Electronic Digital Computer* (1951) became the software bible.

High-Speed Computing Devices had chapters on the basic circuit elements of computing, such as counters and arithmetic elements, "functional approaches" to machine design, and an extensive review of mechanical calculators, punch-card systems, analog computers, and other large-scale computing efforts. Its comprehensive bibliographies of the state of the art in computing were invaluable to engineers at the time and remain of interest to historians today. Even though much of the book is a cautious appraisal of the state of the art, it is worth remembering that people's thinking about "digital" circuits was still in embryonic form. Although the decimal-based ENIAC looked somewhat antiquated, there were several different ways of implementing binary or base-two computing schemes, including raw binary, octal-coded binary, binary-coded decimal (used in Univac I and Univac II) and bi-quinary coded decimal

FIGURE 2.5. Written by staff of Engineering Research Associates, *High-Speed Computing Devices* (1950) was the "Bible" of computer designers for years.

(used in the IBM 650 and several Univac models, including the supercomputer LARC). The ERA engineers even considered an entirely novel base-three or ternary computing scheme, evaluating its efficiency quite favorably in transmitting numerical data as a series of electronic pulses: "A saving of 58 per cent can be gained in going from a binary to a ternary system." Such ternary computers, based on magnetic cores, were actually built by Nikolai P. Brusentsov at Moscow State University between 1958 and 1965.[19]

In 1947–48, ERA landed a contract with CSAW to design, develop, and build the pioneering computer known as Atlas. CSAW arranged the famous Task 13 contract again through the navy's Bureau of Ships. In January 1947, with the various obscure hints from von Neumann and the Moore School suitably arranged, Pendergrass forcefully put forward the case for building a general-purpose computer for cryptography, a document that ERA's research director Arnold Cohen described as "a pretty good sales talk for the utility of a general purpose machine, compared with all of the special hardwired mechanical things. . . . That was a good source of education for me."[20] Once the Atlas project was funded, Cohen worked closely with Pendergrass and other CSAW–NSA staff who were then living in St. Paul to refine the computer's design and specific capabilities.[21] Another NSA staff colleague and Atlas collaborator was Howard Campaigne, who was a math faculty member at the University of Minnesota before the war. "Well, one of the things we wanted to add [was] bit by bit with no carry, that was specified on the first machine," recalled Pendergrass. "But at the time, . . . we were worried about classification. So I think we requested ERA not to put this instruction in." All the same, this classified instruction was indeed included, because Cohen described it as "the exclusive OR . . . on a single bit, which in the crypto trade they were calling false add, and I believe we used the term vector add in the Atlas I." A declassified 1964 NSA history confirms that Atlas I had instructions for both "vector add" and "random jump," which, when coupled with a special piece of hardware, could generate random data strings that were useful in encrypting information.[22] A seemingly random yet reproducible data string would be an electronic means of achieving the holy grail of cryptography, a virtual onetime cryptographic pad that could never be cracked. Cohen, an electrical engineering graduate (1935) of the University of Minnesota, also created a 4-bit prototype known as Cognac to demonstrate the new stored-program concepts. In notes at the time, Cohen described Cognac, short for (rather fancifully) Cogitating Numerical Adder and Computer, as "an inexpensive, impractical hypothetical machine, for illustrative purposes only." Even though digital computers per se were unknown, Cohen's doctoral thesis work under University of Minnesota physicist Al Nier dealt with the nearest thing, much like Mauchly. "I did get experience in building scaling counters, which are binary counters, but we didn't call them binary counters in those days. Things of that nature would later be useful."[23]

Two young veterans of the Goldberg and Demon projects took charge of the engineering of Atlas. John L. "Jack" Hill and Frank Mullaney were fellow ham radio enthusiasts, as it turned out, who met before the war at the St. Paul Radio Club. During the war, each had made the conceptual transition from the continuous waves of radio to the discrete pulses of computing. "I think the concept of a pulse to do something," Mullaney recalled, "the thought of chasing pulses around and opening gates and letting them through . . . in this radar and sonar equipments, certainly had a direct application in computers." Mullaney focused on the control system, including coding and decoding instructions. The "machine instructions . . . came mostly from the customer [NSA]," he recalled. "But what Arnold [Cohen] was doing at that time was taking each instruction and breaking it down into the micro instructions or the steps needed to do it. And then I took it from there and put that into circuitry." Hill worked on the execution of the control signals, including the binary arithmetic unit, which, as he recalled, was "most bothersome to us all because it was so completely foreign to anything we had ever done." Some parts of Atlas were more or less freely borrowed. For example, "the gate/register relationship was copied directly out of [MIT's] Whirlwind," Hill recalled. "We made no attempt at originality; it served no purpose. We were in a hurry; we were encouraged to use anything we could find."[24]

No one was happy with the cumbersome magnetic tape used to wrap ERA's first experimental magnetic drums, where the data bits were physically stored and shuffled. Fortunately enough, there was help ready at hand from one of Minnesota's established industries. ERA's Sidney Rubens quietly pressed a wartime acquaintance at the Minnesota Mining and Manufacturing Company (that is, 3M) for a sample of spray-on magnetic coatings that the company had tested in its research lab but not yet commercialized. Rubens was a Ph.D. physicist who had worked at the Naval Ordnance Laboratory on magnetic mines during the war; he gave a 3M colleague a sample of captured German magnetic recording tape and told him, "when you can make it like this, you will be in business."[25] In return for early field trials of the spray-on magnetic coatings, ERA also did state-of-the-art analysis on using the magnetic surfaces for data storage that helped 3M to refine and commercialize its products. Some modest part of 3M's decades-long global dominance in magnetic tape and magnetic

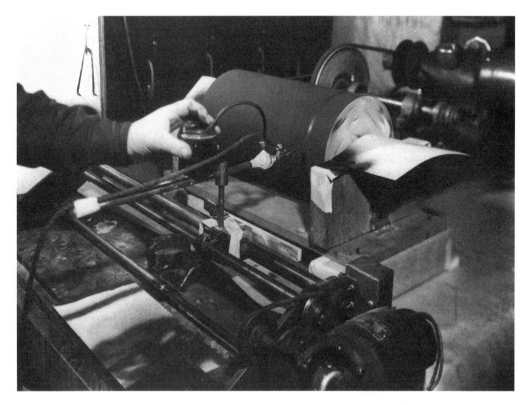

FIGURE 2.6. Manufacturing computer memory drum, 1952. ERA and 3M jointly developed spray-on magnetic oxide coatings for computer memory units.

coatings might fairly be traced to its early interactions with the ERA team. Moreover, 3M is an early instance of an "ancillary industry" usefully contributing its expertise to the nascent Minnesota computer industry (see chapter 7).

The early ERA venture was shaped not only by the industrial district's technical capabilities, including graduates from the university and interactions with 3M, as well as notable electromechanical skills and capabilities built up in the Snelling Shops, but also through tense interactions with the navy. Even though the Atlas computer was always intended strictly for shore duty, the local navy inspectors in St. Paul, evidently interpreting the Bureau of Ships contract in literal terms, scrutinized the machine with standard ship-borne technical specifications. They initially required all electrical wiring for Atlas to conform to a pre–World War I specification—you can find it in a 1911 naval electricians textbook—that called for fat twelve-gauge wiring inside the computer. For a time, the inspectors even insisted that the room-sized Atlas be re-packaged to squeeze down the twenty-four-inch square hatch of a submarine. One inspector rationalized this apparently unreasonable stance because it was possible that Atlas might be transported aboard a navy ship, sometime, and fit through a warship's narrow bulkheads. It was not all bureaucratic blockheadedness. ERA veteran Frank Mullaney allowed that the frighteningly stringent military specifications may have evolved an engineering-design philosophy that resulted in unusual reliability, a notable characteristic of ERA and its successors, including Univac and Control Data, for the next two decades. "It did give us a discipline that we really needed, in that when we started to put together thousands and thousands of tubes and other components that I think we quickly realized that you had to have this kind of conservatism in order to come out with any kind of reliability figure."[26]

Rigid secrecy even *within* ERA was another legacy of the naval-intelligence setting that constrained awareness of the industrial district's growing competences in computing. "We were kind of compartmented. I didn't know what people were doing," stated Mullaney. "We were continually directed to keep to ourselves what we were doing, even from other members of ERA," in Hill's words. "And although we were very, very free in talking to each other within the projects, we just didn't mention what we were doing at lunch in any way that could be revealing as to the real efforts. As a matter of fact, it took me a long, long while to find out that they were working on Williams tubes over in the other room." Williams tubes, a promising alternative memory scheme, were drafted to the cause for the second generation Atlas. And, needless to say, "we were unable to speak to anyone outside of ERA and in our own professional societies about what was going on."[27] ERA was quite literally part of a "closed world" that happened to be square in the Midway district of St. Paul but was otherwise nearly invisible.

Mullaney, Hill, and the other ERA engineers soon completed design, construction, and testing of Atlas, which was a room-sized machine something like ENIAC. Most surviving photographs are of later commercial versions that spotlight a broad U-shaped control console that could sit on a standard government-issue metal desk. A rare photo of Atlas (see Figure 2.7) takes you back. "We had a pretty sophisticated control panel. . . . It looked much more impressive as a computer than today's computers, lots of neon lights," suggested Frank Mullaney.[28] The Atlas computer, with its 2,700 vacuum tubes, 2,400 crystal diodes, and drum memory unit, filled a space 38 feet long and 20 feet wide. The magnetic drum 8½ inches in diameter rotated at 3,500 rpm with 200 read-write heads, looking something like spark plugs sticking out from its cylindrical form, providing read-write access times as quick as 32 microseconds or—if the drum needed to make a full rotation to access the chosen memory location—as slow as 17 milliseconds. For mathematics Atlas had a single 48-bit "accumulator" that recalled the ring counting scheme of ENIAC.

Because the Atlas machine did addition in just ninety-six microseconds, and multiplication a little more than three times that span of time, anyone writing a program for it—just like the legions of IBM 650 programmers to follow—needed to juggle the logical and physical addresses to minimize wait times owing to the spinning magnetic drum. Viewed another way, Atlas could do fifty multiplications in the time it took

FIGURE 2.7. A rare photograph of the pioneering Atlas computer of Engineering Research Associates in 1951. Atlas was installed by December 1950 at the precursor to the National Security Agency in Washington, D.C.

for one drum rotation, but fifty drum rotations took forever. The memory unit was capable of holding 16,384 words each of twenty-four bits in length—something like forty-eight kilobytes today. For this electronic wonder, the best estimate is that CSAW–NSA paid the royal sum of $1 million.[29]

Atlas in the World

In 1950, when Atlas was being readied to begin its journey from the St. Paul factory into the world, it is worth emphasizing that none of the new-generation computers had ever been successfully relocated. Eckert and Mauchly's BINAC successfully ran programs in the spring of 1949—in early April it successfully ran a fifty-line diagnostic program that tested all instructions and checked its memory, clocking that same month an impressive 31½ error-free hours—and as such may be considered the first American stored-program computer. But troubles started in September when it was turned over to Northrop Aircraft Company and then shipped to southern California. BINAC's mercury acoustic-delay memory, invented by Pres Eckert himself, was especially fragile. Each of its twin memory tubes held 512 "words," thirty-one digital bits in length plus a short buffer between words, stored as acoustic pulses traveling down the length of a mercury-filled tube; such tubes were notoriously fickle and needed to be held at a constant perspiration-inducing temperature of 100°F. Very likely, among BINAC's many problems, the mercury delay lines did not make the journey. "It never did operate in California," remembered one participant. "The mercury had apparently deteriorated."[30] In effect, BINAC died somewhere between Philadelphia and Los Angeles.

Other early stored-program computers experienced trauma when it came time to move them. Eckert and Mauchly's UNIVAC computer, also using mercury delay lines for memory, was "delivered" to the Census Bureau in March 1951, but no one dared to physically move it from the Philadelphia shop for nearly two years.[31] With good reason, Eckert worried about the mercury lines: "We were on the second or third time through the [UNIVAC] acceptance test and things were going reasonably well. Somebody decided they were cold in this old building . . . and reached up to close the window. And Pres said, 'Don't touch that! Don't change anything!'"[32] EDVAC, the ENIAC-successor project that stayed at the Moore School, was delivered to the army in August 1949 but suffered debilitating technical problems for two years and did not properly work for an additional year or more. Maurice Wilkes's EDSAC in Cambridge, England, computed a table of squares in May 1949 using mercury delay memory. EDSAC is often cited as the first practical stored-program computer; no one ever dreamed of moving it.

Atlas, with its magnetic-drum memory machined from steel, was physically far more robust than any of these mercury-tube machines. In November 1950, ERA's St. Paul factory was abuzz with excitement. "The individual Atlas cabinets were loaded

FIGURE 2.8. ERA engineers with several sizes of ERA magnetic drums, circa 1955. From left: Jack Hill, Arnold Cohen, Frank Mullaney, Bob Perkins, Arnie Hendrickson, and Bill Keye.

on a [railroad] spur track which came into the ERA premises," recalled Harlan Snyder, a navy lieutenant commander. Snyder had been sent to St. Paul a bit more than a year earlier to learn enough about Atlas to train CSAW staff to use and maintain it. Navy guards accompanied its thousand-mile journey to Washington, D.C., in special Pullman and Railway Express cars. "In Washington it was transferred by truck to Nebraska Avenue, installed, and brought up and running by ERA personnel in less than two weeks."[33] Local lore has it that the engineers sent to Washington—John L. "Jack" Hill and others—were home by Thanksgiving. According to Cohen, "legend has it people were betting on this one . . . whether the ERA installation crew would be home for Christmas. And it was set up and the power turned on and it ran. So, they let it run through various test routines for a week or two, and nothing happened so they went home for Christmas."[34]

There was now one fast general-purpose computing machine in place at Nebraska Avenue, retiring the wartime bombes and special-purpose code machines. Atlas did multiplication at least seven times faster than UNIVAC, and addition more than twenty times faster, and the first UNIVAC was actually yet to be delivered.[35] CSAW was already absorbed into the joint Armed Forces Security Agency and as the postwar reorganization of intelligence proceeded, eventually, in 1952, it became part of the National Security Agency. A second Atlas was delivered to NSA in 1953, and two additional updated models in the following months.[36] These machines were not easy to program; there was no operating system or even assembly language. Because this was even before the invention of "batch" processing, operators lined up at the change of every shift to individually load and run each program. Perhaps looking to best an early EDSAC program that computed a seventy-nine-digit prime number, the largest then known, the Nebraska Avenue group began calculating the digits of the natural constant *pi* to several hundred places. This effort was kept under tight wraps, however, because "there was concern that release of any results could be a security problem as it would raise questions as to how such a result had been calculated, and by whom."[37]

Whatever the satisfying thrill of computing *pi* or prime numbers, the pressures of the Cold War were never far away. "It is believed that the first operational program written for Atlas was designed to attack isologs in VENONA messages," states a declassified NSA history.[38] Venona began in 1943 as an Anglo-American intelligence operation to scrutinize the Soviet Union, and it extended well into the Cold War period. Venona furnished the (then-secret) documentation in several of the most notable spy prosecutions of the 1950s, including the cases against Julius and Ethel Rosenberg, Alger Hiss, Donald Maclean, and Guy Burgess. Venona involved the meticulous matching of mountains of encrypted Soviet messages; owing to the chaos of wartime conditions, the Soviets made the serious blunder of repeatedly reusing their nominally "onetime" cryptographic pads. Venona decryptions continued until 1980, and the operation was made public, finally, in 1995.

Atlas Rebranded

"I remember Bill Norris and Howard Engstrom tripping to Washington to get us to declassify Atlas so they could start to peddle it. Captain [Joseph Wenger], who was the power, was very tight on security. We felt it was an ordinary computer, but he also was concerned at that time about what is our outfit doing spending millions of dollars on computers?" remembered Pendergrass. "But, we did declassify it."[39] ERA hoped a commercial version of its top-secret Atlas would make a mark in the wider world. The differences between the two machines were "very, very minimal," thought Mullaney. "It was taking one [cryptographic] machine instruction out of the repertoire."[40] Echoing the Task 13 contract, Jack Hill proposed that the commercial machine be named "1101," the binary representation for 13—thus mercifully retiring the project's working name of "Mabel." ERA built a limited number of them. Besides the intelligence agency's two copies of Atlas, one 1101 was sent to ERA's research office in Arlington, Virginia, to be the core of a commercial computing service, where clients might rent time for doing calculations, but this activity never really took off—"it was way too ahead of its time," remembered one staffer—and the machine was donated to Georgia Tech a few years later.[41] Three copies of a closely related 1102 model were sent to the air force's Arnold Engineering Development Center for analyzing wind-tunnel data. It was with the nineteen copies of the revamped 1103 model, fitted with those secrecy-shrouded Williams tubes, that ERA finally found a measure of success. The navy took delivery of the first two copies of the 1103 for nearly $2 million, while the third was rented to Convair, the Southern California aerospace concern.

Although it sounds odd today given Silicon Valley's inescapable pull, the Convair installation occasioned a significant migration from California to Minnesota. Convair (soon to become a division of General Dynamics) was a classic Southern California aerospace concern. In these years, fully fifteen of the nation's twenty-five largest aerospace companies were located in Southern California, including the legendary Douglas, Rockwell, North American, and Hughes aircraft and aerospace concerns, along with RAND in Santa Monica and the Jet Propulsion Laboratory in Pasadena. They each had an insatiable appetite for computing and the deep pockets, owing to

FIGURE 2.9. This "ERA 1103" computer was used for analyzing wind-tunnel data at NASA Lewis Research Center in Cleveland. ERA computers were extensively used by military agencies and aerospace companies. Source: Weik, *A Third Survey of Domestic Electronic Digital Computing Systems*, 906.

governmental contracting, to pay for it. A young Robert Price, later CEO of Control Data, cut his programmer's teeth on Convair's ERA 1103; his boss Ben Ferber later came to Control Data as well. Also posted at Convair for a time was ERA's Erwin Tomash, who talked up the Twin Cities' burgeoning computer scene to Marvin Stein, a recent UCLA math doctorate, who before long would take up a position at the University of Minnesota, running the university's new computing center built around an 1103 (discussed later in this chapter).[42] The ERA 1103 would be best known, however, as the Univac 1103.

ERA, like the Eckert and Mauchly company before it, lost its independent existence in a cloud of suspicious politics and unsteady finances. John Parker's original financial structure from 1946, designed to support a research-oriented company, was by any reckoning woefully insufficient for a production-oriented company that needed factory space, expensive supplies, and a sizable labor force. "We would rush down to the bank every Friday to borrow on the invoices to the Navy to cover the checks that had already been written," admitted ERA's treasurer. "We got to the point that if we were going to go commercial . . . on this business . . . we needed lots of money," Parker recalled.[43] He made plans to somehow associate ERA with a larger company, because ERA's top-secret cryptography work made it impossible to reorganize ERA as a normal publicly traded company. "I did my best to acquire some business from IBM," he stated. "Several times we came down to almost having perhaps a major contractual product arrangement with IBM."[44]

Parker certainly made some unusual and generous overtures, which typically began with his inviting some business colleagues in for a visit to show off the facilities at ERA. "I remember a series of meetings with Honeywell, management and technical people. . . . What was really in mind was acquisition by Honeywell of ERA," stated one participant in the parade. Perhaps the most far-reaching overture was an ERA study contract to assess and improve IBM's magnetic-drum technology that resulted in IBM's gaining ownership of "two fairly massive patent specifications" that paved the way for its own Magnetic Drum Calculator, which as the IBM 650 sold a phenomenal two thousand units over its nearly decade-long lifetime.[45] Yet IBM declined anything further. Earlier, it so happened, Eckert and Mauchly had tried to sell their company to IBM, but also to no avail.

In this unsettled time, the unusual circumstances of ERA's founding came to light. Drew Pearson's nationally syndicated "Washington Merry-Go-Round" was closely read in the capital for its leaked secrets and insider innuendo. Normally, Pearson went after big fish. His persistent attacks on James Forrestal, delivered in his daily newspaper column and weekly radio address that spread "the corrosive personal abuse of gossips and keyhole commentators" quite possibly contributed to the defense secretary's suicide in May 1949, two months after Truman relieved him from the post. Pearson then energetically stirred the pot against General Douglas MacArthur, both before and after war broke out in Korea. Pearson's deputy muckraker Jack Anderson likely heard the ERA story from his wartime buddies in the Office of Strategic Services (precursor to the CIA), though Anderson was also unusually friendly with Senator Joseph McCarthy. "Navy officers find cushy jobs with company awarded contract," was the headline for the August 16, 1950 column. The article's mix of half-baked speculation and mundane detail bestowed a kiss of death:

> One of the Navy's most closely guarded secrets is a project that had priority over the atomic bomb during the war. It is still considered so secret that we cannot disclose what it is. However, we can reveal that the Navy entrusted this vital project, involving complex engineering and construction, to an inexperienced company. Later the same Navy officers who had made the deal turned up as highly salaried vice presidents of the company. The outfit that wangled this highly secret, multimillion-dollar contract was Northwest Aeronautical Corporation, later reorganized as Engineering Research Associates of St. Paul, Minn.

Singled out to blame for this "juicy contract" were the ERA vice presidents Ralph Meader, Howard Engstrom, and William Norris. Meader's recent cashing out for thirty thousand dollars with a "mysterious clause" prohibiting any legal claims on the company added a frisson of scandal.[46]

Pearson's muckraking account of this "vital project" appeared in perhaps three hundred newspapers across the country. The public-relations angle was threatening enough that Parker, hearing the news on his plane's radio while flying home from Washington, immediately on landing in St. Paul sent word to an Annapolis classmate, high up in naval intelligence, and "got the story squelched in about 200 odd papers." The word was out all the same. Who could deny that a unique business opportunity had landed in the laps of Meader, Engstrom, and Norris? Indeed, that these former navy officers had themselves actively created ERA? Even, that they were handsomely funded by their former navy boss, none other than the intelligence vice chief Joseph Wenger? "It looked like Wenger was just feeding his favorites and [his rivals in the navy] didn't like it. So somebody came along and said it's time that ERA found some other business," thought one insider. Pearson soon enough trained his sights on other Washington targets, but for ERA the damage was done. "The relationship with the Navy's procurement offices [became] so strained that Joseph Wenger was unable to get an official letter of thanks sent to ERA for having developed the Atlas."[47]

ERA was about to lose its independent existence. Remington Rand, having already bought up the Eckert and Mauchly company, took a hard look at the ERA company and drove a hard bargain with Parker. "John Parker the capitalist [was] courted by people who wanted to buy something he owned and controlled. And he did control it in terms of ownership. . . . And he wanted to make the sale; actually, nobody else

FIGURE 2.10. The office-machines business of Remington Rand, before 1940. The woman at right has a Remington typewriter. ERA's engineers and managers often mocked Remington Rand as "the typewriter people."

in ERA wanted to make the sale," according to one ERA veteran. "We could be employees of someone we didn't even know; shavers and typewriters were not our bag. So it turned out to be a very bitter fight internally." "There was a feeling on the part of the ERA people that they [i.e., Remington Rand] were rather inept—my god if we let them in they'll screw it up," recalled a second. Despite the early founding vision that led Norris and the engineer associates to value their independence, the company was Parker's and, given his ownership of fully half the outstanding shares, his alone to sell. Parker initially pegged ERA's valuation at $5,000 for each of the company's employees, but, after the tough negotiation with Remington Rand was done, they were each worth only $1,400 in a stock swap.[48]

Remington Rand's critical appraisal of ERA constituted the opening salvo in what eventually emerged as a pitched corporate rivalry between St. Paul and Philadelphia. At the time, ERA, at least as Remington Rand saw it, was "predominantly of an engineering character," employing 158 engineers, sixty draftsmen, and additional technicians, but without significant production facilities. Indeed, "much of their construction technique is quite similar to that employed at Philadelphia and is of a type which will infringe patents now allowed and scheduled to issue within the next nine months," wrote Remington Rand's financial adviser. He happened to be a proverbial Philadelphia lawyer—none other than Eckert and Mauchly's chief legal counsel.[49]

With its Philadelphia and St. Paul acquisitions completed by early 1952, Remington Rand owned the world's two leading commercial computer companies. The company's president, James Rand, reportedly thought that "if he owned Eckert-Mauchly and he owned ERA he would own 95 percent of what there was and he could take their expertise about how to manufacture products and sell them to businesses and beat IBM."[50] Remington Rand also had a tabulating machine factory in Norwalk, Connecticut, and it might have knit these promising elements into a true computing colossus. For a time, especially with the famous televised demonstration the night of the 1952 U.S. presidential election, "Univac" was even the generic term for a computer, much like Coke or Levi's. One of the Univacs was installed at General Electric's showcase automated appliance factory at Louisville, Kentucky, the first American commercial computer user in 1954.[51] George Gray identifies several promising lines of complementary technologies, including Philadelphia's magnetic tape drives and St. Paul's magnetic drums that were shared between the two divisions.

Still, to put it mildly, "there was quite a large disparity between the electronic design philosophy of the two development laboratories, Philadelphia and St. Paul." "There was a great deal of locational rivalry between Eckert–Mauchly and ERA for attention on the part of Remington Rand management and for funds, allocation of funds, for research and development," stated one insider.[52] "I think they had a more intellectual and analytical approach," observed ERA's Mullaney, in a generous and balanced assessment. "They were also very innovative, but I think we built better hardware. Consequently, we looked at ourselves as the practical people that could make things work and we looked at them as these eastern fellows that thought they were pretty hot stuff. And it just didn't work very well."[53] Relations between the rival divisions went from bad to worse. An early visit to St. Paul by Remington Rand's research-and-development director Leslie Groves—the infamously blunt former army general, builder of the Pentagon, and wartime Manhattan Project chief—went straight south. Groves, as one ERA veteran recalled it, announced, "Well, it's obvious we've got in Remington-Rand now, two competing computer systems . . . we obviously can't afford this. I would like all of you to go into the next room and come back in about 15 minutes and tell us how we can make one product line out of it."

When the Philadelphia and St. Paul groups could reach no agreement, he ordered them to repeat the unpromising exercise—then threatened to fire them all. Sales and financing were additional friction points. Rand himself never bothered to visit St. Paul.[54] And yet, despite these notable differences and rough spots, in historian Arthur Norberg's assessment, "integration of ERA into Remington Rand seemed to be going smoothly."[55]

Early on, there was a chance for ERA to maintain a distinct and valued identity within the parent company. ERA "was generating good revenue and equally good profits in a somewhat different field, scientific computation, which the Remington Rand management had virtually no feel for," one observer noted. There was certainly room for accommodation. Then came the merger in 1955 between Remington Rand and Sperry. "Sperry was quite clearly manned with many top level management executive personnel quite knowledgeable about technology, see they had the experience in the automatic pilot and radar work and microwave work from World War II."[56] Another ERA veteran observed that "We had a little bit higher regard for the Sperry people, because we felt that they were more our kind of guys, I guess. . . . old man [Harry] Vickers . . . was kind of an old shoe type engineer."[57] "Harry Vickers understood technology; he was a ham [radio operator] and also a creative engineer and had had a great deal of experience dealing with the Sperry people in that business of war so you know a whole new cadre of top executive management was coming into place," thought another.[58]

Yet, in 1955, when Sperry bought up Remington Rand and ERA was absorbed into the Univac division of the merged corporation, the Philadelphians scored another victory, symbolic but important. Ever since, it has rankled ERA veterans to have their company effectively rebranded as an appendage of the Philadelphia story. In midstream, the ERA 1103 became the Univac 1103 or Univac Scientific computer; and some accounts even go so far as to retrospectively relabel ERA's original commercial computer as the Univac 1101. It certainly would have altered the computing identity of Minnesota if, hypothetically, Sperry Rand's Univac division that came to play an immense role in the state—with eight major factories and R&D facilities in the Twin Cities and a string of nationally important technical achievements (see chapter 3)—had instead been named the "ERA division."

When the University of Minnesota came to installing its first digital computer, "ERA" had already been effaced from the scene. This turn of affairs was unfortunate because the university had provided much of the brainpower behind the ERA venture, as already noted. Among the designers of the Univac–ERA 1103 were Seymour Cray (electrical engineering class of 1949 and applied math in 1951) and Arnold Cohen (electrical engineering class of 1935 and one of the numerous Ph.D. students of Al Nier's in physics), who was later associate dean of the university's engineering school. When Univac heard rumors that the university was considering acquisition of

FIGURE 2.11. ERA 1103 relabeled as "Univac Scientific" computer, 1958. From left: University of Minnesota president J. L. Morrill, Marcell Rand, electrical engineer W. G. (Jerry) Shepherd, and Univac's Kenneth P. Pitterman at the unveiling of the computer at the University of Minnesota.

an IBM 650, it presented the university with an irresistible offer of four hundred hours of computer time on a Univac 1103. According to Marvin Stein, who was hired for the position running the new computer center in 1955, the intent was in part to put the machine to useful work as well as to develop courses that "viewed [computers] as objects of intrinsic interest." With a Ph.D. from UCLA, Stein already had experience with the 1103 computer at Convair, the prominent Southern California aerospace concern. It so happened that Erwin Tomash (electrical engineering class of 1943) and other high-level ERA–Univac staff worked with Stein on the installation at Convair, and so connected these California programmers to the computing scene in Minnesota. "On a Friday afternoon I paid my visit to the IT Dean's office and spoke with Gerry Shepherd, and on Monday I had an offer. Things could really move quite quickly back in those days," he recalled.[59]

The four hundred hours of computer time were quickly taken up by researchers across the university. It's impossible to review these early computing projects and not sense that something big was afoot. Chemistry professor William N. Lipscomb was the largest single computer user during the academic years 1955–57, analyzing data on chemical bonding for which (after leaving for Harvard) he eventually won the 1976 Nobel Prize in chemistry. Neal Amundson and Rutherford Aris used computing to reinvent chemical engineering (an early expression was Aris's textbook *The Optimal Design of Chemical Reactors: A Study in Dynamic Programming* [1961]). Reformulated around advanced mathematics and computing, chemical engineering at the university became the nationally ranked number one department for decades. Ernst Eckert worked a similar transformation in mechanical engineering, using extensive computing power to build up an internationally regarded profile in the new specialty of heat transfer. Computing time shaped the landmark Minnesota Multiphasic Personality Inventory, or MMPI personality test. An early research project in statistical analysis and econometrics involved Leonid Hurwicz, another future Nobel Laureate (see chapter 7). Indeed, the campus computing demand was so strong that researchers in the physics department even took one of their computational projects to Convair's 1103 in California.[60]

Soon the new year-long computing course was attracting "an extraordinary amount of interest," and there were far more computing projects requesting time than the generous Univac grant could possibly cover. The university raised funds and in December 1957 finally made an outright purchase of a "Univac Scientific" computer for just $250,000. Univac awarded the university a bargain-basement price on an expensive state-of-the-art machine, installed at the university with a full squad of trained programmers and technicians, but the fine print of the contract gave the company free use of the computer for up to *eighty hours* a week (entirely taking over the night shift from 5 p.m. to 8 a.m.) for two years. Stein, as director of the university computer center, rapidly expanded the university's computing muscle with three Control Data machines in the 1960s and subsequently became a key figure in establishing computer science as a full-fledged department at the university in 1970.[61] Pretty clearly, the state's computing industry and the state's research university were joined at the hip.

3
Corporate Computing

Univac Creates a High-Tech Minnesota Industry

Minnesota's computing destiny, during much of the 1950s, was decided at Remington Rand's corporate headquarters in Connecticut. Already before the war, James Rand Jr. had successfully stitched together the typewriter and firearm branches of his family's businesses, all the while keeping a weather eye out for fame and publicity. When a granite-faced Tudor mansion originally built by a U.S. Steel president came onto the market in the old-moneyed part of Norwalk, Connecticut, he snapped it up in 1943. There he installed his company's executives and made expansive plans for the postwar economy. The first step was hiring famous generals. Rand secured the services of army general Leslie Groves, the recently retired builder of the Pentagon and commander of the Manhattan Engineering District, to be head of research and development in 1948. So far as anyone can tell, Groves in the subsequent years spent less time and effort at the brand-new laboratory in South Norwalk than on traveling around the country giving speeches and burnishing his place in history as the builder of the atomic bomb.[1] As chapter 2 recounted, Groves was not loved in St. Paul.

To fill out his table at the Rock Ledge mansion, Rand named another recently retired army general as chairman of the board in 1952, around the time of the ERA acquisition when John Parker was an occasional guest. From then on, Douglas MacArthur also graced the weekly Thursday luncheons. These were events not to be missed. Rand loved to spark freewheeling rounds of "lunch-table inventing" where the assembled dignitaries dreamed about the next generation of new computers, though it seems they had little regard for the actual capabilities of the company's engineers in Philadelphia or St. Paul or even the tabulating-machine factory just down the road. In a pause in the conversation after lunch, someone would invariably ask MacArthur what he thought about the latest idea for a technical breakthrough.

"He'd take a couple of drags on his corn cob pipe and he'd say, 'Well, that reminds me of the time I was . . .' and then he'd tell some sort of war anecdote not having anything to do with the question, but that was his charm."[2] One St. Paul man suggested that "MacArthur was very significant in terms of his contribution to the Univac image [especially] in Japan," while another executive thought "he was magnificent" in entertaining top executives to clinch sales.[3]

The 1955 merger of parent Remington Rand with Sperry, a company long known for sophisticated autopilots, gyrocompasses, bombsights, and other military-related precision technologies, might have set the computing world on fire. For a brief time, Sperry Rand Univac was in a singular position, at least until IBM trained its formidable sights on the computer market. Even so, Univac made the Twin Cities into a computing powerhouse, soon employing ten thousand Minnesotans and anchoring the state's emerging computer industry. Spinoffs from Univac such as Control Data, treated in the next chapter, added to the region's computing muscle and created the country's first industrial district specializing in computing. And, notably,

FIGURE 3.1. Remington Rand's "Rock Ledge" headquarters, 2007. This mansion was the site of James Rand's "lunch-table inventing" with retired army generals in the 1950s. For years Minnesota's computing future was decided here in Connecticut. Source: Noroton permission by GNU Free Documentation License, version 1.2; Creative Commons Attribution-ShareAlike 3.0 License.

from Univac's own Minnesota laboratories and factories came new computers that reliably networked the navy's battleships and safely guided the nation's commercial and military aircraft. What is more, Univac's assertive and determined efforts to whip into shape its suppliers in the semiconductor industry worked an early "quality revolution" that, as we can now appreciate, paved the way for "Moore's Law" to kick in. Univac's contributions to Minnesota computing, while grounded in the Twin Cities, played out on a national stage.

Thousand-Mile Abyss

With the Sperry Rand merger in 1955, Bill Norris was assigned the unenviable task of wrestling the competing units of the Univac computing division into some form of cooperation. From the start, it was tough going for Norris, the former ERA vice president and all-around troubleshooter. "Pres Eckert took the view that what ERA was doing was not state of the art. Therefore, he didn't want to waste his time with us . . . it was terrible," Norris recalled of the early Rem Rand years. Norris as general manager of the Univac division had authority over both Philadelphia and St. Paul as well as top-level access to the company's executives, including the Thursday luncheons. The needed elements seemed to be in place to mount a significant and successful effort to stay well ahead of IBM. But, in the wake of the Sperry merger, deadly corporate infighting emerged when the manufacturing vice president aimed at gaining control also over the engineering staff. Effective cooperation between the manufacturing and engineering divisions was needed—certainly not a marriage, but neither a divorce. Norris was caught in a trap. A corporate plan to entirely separate engineering from manufacturing, circulating in early 1957 "was just goddamn ignorance on the part of the top management of Sperry Rand," he thought.[4] Norris also opposed a misbegotten attempt to impose a wage cut on the St. Paul workforce by bribing local union officials. After giving it the good fight, Norris decided to cash out and try again at the start-up game. "The environment [was] ripe for something new to happen, and Arnold Ryden came along with a plan, just at the right time. He was able to interest Bill Norris and others to join with him in . . . the forming of Control Data."[5] Norris joined Control Data Corporation that summer and presided over its swift ascent into the top ranks of computing (see chapter 4).

The impression that they were second stringers in the Sperry Rand sphere deepened when St. Paul was assigned the development and engineering work on Univac II. Pres Eckert, the company's anointed engineering genius, "a very strong person in terms of his ability to articulate," had reeled in the deep pockets of the Lawrence Livermore national laboratory and started work in Philadelphia on a high-end transistor-based computer—the resulting $6 million Livermore Advanced Research Computer, or LARC, was briefly the fastest computer in the world until the even more expensive IBM 7030 Stretch came along—and he couldn't be bothered any longer with prosaic

74 CORPORATE COMPUTING

commercial machines.⁶ St. Paul took the hit. "Larc was going so far in the hole that it soaked up all R&D dollars and there wasn't anything left over to start anything new. In fact, all budgets were slashed in St. Paul," recalled one engineer. Another remembered that ERA veteran "Jack Hill . . . took an increasing amount of heat from Pres Eckert personally on this particular project, and finally had had a belly full of the whole thing and left the company in 1956, so we have a time marker on that program."⁷ The original Univac had been a high-profile, state-of-the art computer that sold an impressive forty-eight systems to leading companies and prominent government agencies. But Univac II looked to be an uninspired bet on a mule. It relied on an ungainly hybrid technology, with no substantial changes besides the retirement of the troublesome mercury delay lines by magnetic core memory and sundry upgrades of storage and peripherals. "The development of Univac II should not have been undertaken at all because at best it would not be a competitive system compared with what was already available," thought one participant.⁸

From the moment of its creation, Univac II suffered from the high cost of its 1,200 temperamental transistors as well as the heat and bulk of its 5,200 vacuum tubes. Designed specifically to run legacy programs written for Univac I, it had the same

FIGURE 3.2. St. Paul was assigned the engineering for the second-generation Univac II, an expensive transistor-tube hybrid that failed in the commercial market. Here is an installation at the U.S. Navy Electronics Supply Office in Great Lakes, Illinois, 1961. Source: Weik, *A Third Survey of Domestic Electronic Digital Computing Systems*, 992.

logical structure, the same mixed binary-coded-decimal number scheme, and the same vacuum-tube-based arithmetic unit. Univac II's addition, multiplication, and division speeds, previously impressive, no longer looked even halfway respectable. It was once good newspaper copy to point out the massive size of your computers, but at twenty-six tons, nearly twice the weight of the original Univac, and a base price of $970,000, Univac II was an overweight and expensive dinosaur.[9] It also was plagued by numerous difficulties in design and testing owing to the "thousand-mile Philadelphia–St. Paul abyss" that hindered the transfer of many essential engineering details. After innumerable frustrating delays with the thirty-three-person test team in St. Paul working three shifts around the clock, an engineering team arrived from Philadelphia (supposedly) to save the day. "The Philadelphia invasion caused about as much resentment as you would expect," recalled one St. Paul engineer.[10] Sales of Univac II, apart from the first five that were preordered and kept the pressure on for timely deliveries, trickled in slowly. When sales petered out at a disappointing twenty-nine units (1958–60), the model was summarily withdrawn.

Manufacturing Machines

In both Philadelphia and St. Paul, it took years to achieve the difficult transition from conducting government-funded research on computing to actually manufacturing salable computers. Bill Norris understood, better than most, that what Sperry Rand had bought into was not so much a guaranteed means to enter the computer industry but rather, as he phrased it, "a chance to get into the computer business by investing a hell of a lot more in R&D."[11] The ERA–Univac 1103 was the test case for manufacturing in St. Paul. Originally, Rem Rand's executives could not have dreamed that ERA had a second commercial machine ready for the market besides the original Atlas-derived 1101. Atlas II was top secret, after all, and so was the existence of the 1103 resulting from it. Engineers from St. Paul sprang the surprise at the Norwalk, Connecticut, laboratory, where "we had some large flip charts, black on yellow paper . . . [with] side-by-side comparisons . . . of the IBM 701 versus the proposed ERA 1103. There was a heavy bullet showing which column was superior on each of those items. That was very impressive." And so was the audience's response, as it turned out: "from where I was sitting I could see Mr. Rand almost in profile, and when Erwin [Tomash] was showing those flipcharts and the point-by-point comparison, Rand moved further and further forward on his chair and then took out his pocket glasses so he could really see that chart better. He was the picture of total concentration."[12]

With the Connecticut corporate go-ahead secured, setting up manufacturing facilities for the 1103 was an entirely new and challenging undertaking. "Production was rather crude at ERA," one engineer noted of the earliest machines. "I can remember many times running down to the assembly shop with a red pencil and marking up a diagram and changing things right on the spot."[13] To undertake manufacturing of the

1103 model, managers in St. Paul had to schedule deliveries of thousands of components and coordinate the work of hundreds of assembly workers. Over the years, the 1103 was re-branded under the new corporate regime as the Univac Scientific to distinguish it from the business-oriented Univacs coming from the Philadelphia-area shops. A total of nineteen or twenty orders for the 1103 model generated handsome profits because the price tag on each machine was just shy of $1 million, comparable to the Univac II, and the engineering development work was already paid for thanks to the Atlas II contracts. The two machines, while aimed at entirely different markets, were in at least one instance too close for comfort. Sperry Rand executives issued a stern corporate rebuke when one ERA engineer made an overly enthusiastic sales pitch, crowing about the superior performance of the 1103 to a Southern California airplane manufacturer.[14]

Of course, more than hardware was necessary to put usable computing power into the hands of users, as IBM had already mastered with its decades of sales and service in the tabulating and business machine world (see chapter 6). Certainly, the St. Paul efforts at pricing, rentals, and customer training were conspicuously weak in comparison. Two ERA insiders even acknowledged the absence of software applications and field support was a "glaring deficiency."[15] Perhaps the true missed opportunity in these early days of the computer industry was not so much in integrating the technical designs of Philadelphia and St. Paul but rather in failing to bring Sperry Rand's substantial experience with marketing and after-sales support to the well-engineered Minnesota machines.

St. Paul's first really successful manufacturing venture came as an offshoot from a 1949 air force contract. The air force needed a compact and rugged device that would automatically adjust, or tune, the electrical properties of an aircraft's antenna to match the radio's circuitry and frequencies. Not only did the device need to fit into a small and exposed area in the aircraft's tail top or wingtip, withstanding extreme temperatures down to minus 65°F, but it also needed to achieve accurate tuning within ten seconds. The St. Paul engineers successfully designed such an "antenna coupler," first manufactured in 1953, and it rapidly became standard equipment for the air force and many aircraft manufacturers. With up to one hundred antenna couplers coming off the production line each month, as Norris recalled, "it was a pretty nice business."[16] In 1957, when antenna couplers generated 38 percent of the entire Univac St. Paul branch's sales volume, and fully 88 percent of its profits, it was understood that when "the new plant was opened on West 7th Street . . . it wouldn't have been built but for the coupler profits." Over a long and profitable run of seventeen years (1953–70), some twelve thousand antenna couplers were sold to Boeing, Lockheed, General Dynamics, and the air force to be installed in scores of military and commercial aircraft.[17] Looking back, the manufacturing head of the Univac plant observed that "this field of communications equipment . . . was high volume production and it was able to carry a lot of overhead and throw off a lot of cash flow and provide some profit, and if it hadn't been for that program ERA, I think, would have died."[18]

Groundbreaking for the $2 million West Seventh Street plant in St. Paul explicitly invoked the state's pioneer heritage. "Historic Shovel to Break Ground," announced the *St. Paul Pioneer Press* in November 1955. The seventeen-acre parcel of land for Univac's first major expansion in the Twin Cities was originally part of the one hundred thousand acres that explorer Zebulon Pike bought from the local Dakota tribes in 1805. Pike Island, named for the explorer, was just beneath Fort Snelling, where in 1819 white settlers had established an early and permanent presence in the state, prior to the founding of St. Paul two decades later. All of these historical details, including the nearby location of the Henry Sibley House belonging to Minnesota's first governor, were lovingly related. Then came the pithy news that Univac's William C. Norris would use the large, square wooden shovel, which had been "dug up" by the Chamber of Commerce at the Minnesota Historical Society, to break ground for the new computer plant.[19]

Univac's new West Seventh Street plant was a distinctive urban facility. The location was clearly within the urban core and not far from the two existing ERA plants on Minnehaha and University avenues (there was also a third smaller plant in downtown St. Paul at 543 St. Peter). A building complex containing more than two hundred thousand square feet of office, laboratory, and factory space was placed on a compact seventeen-acre plot of ground next to the Mississippi River bluff. While other blocks in the surrounding area had filled up with residential housing, this parcel was rock-capped and faced unattractive expenses for residential sewer connections. The St. Paul city council quickly approved the needed rezoning, eager to accommodate the company's expanded payroll of $22 million, which was comparable in size to the large payrolls at Ford, 3M, and the massive Brown and Bigelow publishing company. Most notably, although Univac planned to employ three thousand or more new workers at the West Seventh Street plant, it made provisions for parking only 1,350 cars.[20] A dozen years later, when it moved to the suburbs with its Univac Park in Eagan, the company planned one parking space for each and every employee.

Although we think of technology and engineering as a "man's world," especially in the 1950s, women played a large and visible role at ERA–Univac. A news article relayed the possibly unsettling news that "Feminine Invasion of Assembly Shops Successful." The unionized assembly force had been "an all-male stronghold" until the middle of 1951 when expansion in ERA's assembly volume—soon augmented by the booming antenna coupler business—ran into a serious labor shortage that made it difficult to hire additional male workers. In June that year the International Association of Machinists signed a new contract dividing its workforce into two segments, opening up a second grade of electrical assembler where women flooded in at lower rates of pay. Many men moved up into the ranks of inspectors and foremen, while

FIGURE 3.3. Women at ERA assembling antenna couplers, circa 1955. Antenna couplers brought hefty profits and valuable manufacturing experience to St. Paul.

others transferred into other lines of assembly work. Capable women moved through the ranks starting from trainee bench assembler up to detail assembler, where "she finds herself working from prints with less supervision on non-repetitive work." (It is not clear that women ever moved up further to the rank of inspector or foreman; see the gender-specific employment patterns in chapter 7.) Soon the two assembly shops had thirty-four men and 101 women. At ERA's plant no. 3 on University Avenue near Raymond, thirty women were the principal assembly workforce for antenna couplers. Even in engineering, where the slide-rule-toting male stereotype was taking shape, women were also surprisingly prominent. In the ERA drafting room, there

FIGURE 3.4. Women and men in drafting room at Univac–ERA, 1950s. At the front table are Dick Swifka *(left)* and Sid Burton. Five women are visible here, plus another woman in the original photograph at far right not seen here.

were six women and nine men—a far greater proportion than the roughly 10 percent of engineering and computer science graduates who are women today.[21]

Networking the Navy

The Naval Tactical Data System, or NTDS, was the most significant military system that St. Paul's Univac division ever built, but it was far from the largest military computer project in the late 1950s. Nothing, really, could ever top the mammoth Project SAGE, or Semi-Automatic Ground Environment, the famous follow-on to the Whirlwind computer built at MIT by Jay Forrester. When the navy, which had originally funded the development of Whirlwind as a flight-training simulator, grew wary of his soaring expenses and unbounded ambitions, Forrester subsequently sold the air

80 CORPORATE COMPUTING

force on a grandiose computerized air-defense system that was intended to locate, identify, track, and target incoming Soviet bombers. Some wags even whispered that Forrester was aiming for a mega-effort something like a second Manhattan Project. Without troublesome budget restraint, Forrester designed a top-of-the line computer, and then IBM manufactured fifty-two copies of it. SAGE was truly an impressive heavyweight: each (duplexed) computer weighed 275 tons, and its 55,000 vacuum tubes consumed 1.5 megawatts of power and required a sizable 150 foot by 150 foot room.[22] Programming eventually stretched to a half million lines of code, very likely the largest computer program in the world at the time, while the "effective operating rate is about 75,000 instructions per second."[23]

SAGE was so big that it is difficult to get a grasp of its entirety. MIT's Lincoln Laboratory did much of the conceptual engineering, splitting off the immense jobs of programming to the System Development Corporation and systems integration to MITRE, two completely new corporate entities that would have notable profiles in the years to come. IBM's manufacturing contract was worth $500 million; at its peak, seven thousand IBM employees in Poughkeepsie, Kingston, and Cambridge were devoted to the project.[24] The computer was the brains of the twenty-three SAGE direction centers and three control centers, each housed in a four-story windowless concrete building. The computer itself took up an entire floor of 22,500 square feet, and the necessary air-conditioning and communications equipment, including the linkages for incoming radar, took up another floor. Air force personnel

FIGURE 3.5. The mammoth SAGE computer (circa 1961) required fifty-five thousand vacuum tubes. Univac engineers built a transistorized substitute for it with the NTDS system. Source: Weik, *A Third Survey of Domestic Electronic Digital Computing Systems*, 41.

peering into radar screens for mapping, surveillance, and weapons direction occupied the top floor, while the third floor housed offices and the oversize projector for the two-story "air-situation display screen."[25] Estimates on the total price tag for SAGE begin at $8 *billion* (1964 dollars). The first SAGE center was scanning the northern skies for Soviet bombers in 1959, but by the time the SAGE system was completed four years later, the Soviets had switched to fast-flying ICBMs—rendering obsolete the comparatively slow aircraft detection of SAGE. Ironically enough, by the time it was finally withdrawn from service in 1984, SAGE was entirely dependent on replacement vacuum tubes manufactured in the Warsaw bloc.[26] Owing something to its high profile at MIT and IBM, there are at last count six book-length treatments of SAGE.[27]

Nearly nothing about the navy's own version of the air force's SAGE fits this outsize tale of big iron. Yet NTDS was in its way every bit as impressive, just as fast, and much smaller than SAGE. Its modest cost was also a fraction of SAGE's. Seymour Cray at Univac in St. Paul designed a prototype computer as the "brains" of NTDS. It was an early transistorized computer, in a period when transistors were just emerging from their status as laboratory curiosities, requiring shrewd choices and careful judgment about the rapidly changing technology. Some accounts place Cray at the

FIGURE 3.6. SAGE Direction Center at McGuire Air Force Base, circa 1958, the first of twenty-three similar installations. The SAGE computer took up the entire second floor of the square windowless building. Source: U.S. Air Force.

center of NTDS, but in reality the six copies of his NTDS prototype computer served only for testing the system concept; the production version of the computer was entirely redesigned after he had departed for Control Data. It was this redesigned production computer that became the heart of the navy's battle-fleet communications. Indeed, not only Cray but many other of the key people in the lengthy NTDS effort moved on sometime during the course of the project. More than anything else, an intense and driven partnership between the navy's NTDS project office in San Diego and the Sperry Univac division in St. Paul propelled the system forward from prototyping in the late 1950s to full fleet deployment in the mid-1960s.[28] These relationships with the navy, and especially a hard-won reputation for achieving unheard-of levels of reliability, solidified the position of the Univac division—and Minnesota's burgeoning high-technology economy—as the navy's go-to place for high-reliability computers for decades to come.

SAGE served as the programmatic prompt for NTDS, but the oversize vacuum-tube computer was nothing like a direct model. What became NTDS was first sketched out by an interservice study committee in 1954, when SAGE was already launched as a large-scale bomber defense project. Irvin McNally, attached to the Naval Electronics Laboratory in San Diego (soon to become the NTDS project office) became frustrated after month after month of technical briefings that kept coming around to the pointed suggestion that the navy should adopt a centralized computing model patterned on SAGE. The study committee's illustrious chairman, MIT's Jerrold Zacharias, was up to his eyeballs in SAGE and inclined to see it as a solution for all problems. The navy, as McNally knew, simply could never work with such a centralized model for tactical communication. If a core communications ship were sunk or damaged the entire fleet would be cut off from tactical communication—rendering the ships effectively blind in the midst of battle. Sometimes, even, a single ship needed to engage enemy forces by itself. And anything remotely as big as a SAGE-sized building, requiring a dedicated power supply of 1.5 megawatts, was never going to fit onto a warship. Already McNally was inclining toward much smaller computers using transistors. McNally received some encouragement from Nathaniel Rochester, one of IBM's famed computer designers, who backed his design suggestions and had them included in the final study group report of March 1955, despite the reservations of chairman Zacharias.[29]

The Minnesota network of navy officers and computer engineers decisively shaped the evolution of NTDS when Edward Svendsen teamed up with McNally to draft the "operations requirements" document that would launch NTDS. A native of the state, like McNally, Svendsen had attended the University of Minnesota for a year and then gained wartime experience as a radar officer aboard the battleship *Mississippi*. He next took an advanced course in radar at the Naval Postgraduate School, then still in Annapolis. In 1947, he was assigned to the just-relocated Naval Computing

Machinery Laboratory in St. Paul, where he took charge of the cryptographic computing work, including participating in the design of ERA's Atlas computers. Just possibly Svendsen was the navy inspector who tangled with ERA engineer Jack Hill about whether Atlas should fit down the hatch of a submarine. McNally and Svendsen crafted the technical specification for the NTDS computer, conceived as the shipboard electronic brains to coordinate antiaircraft attacks. They also sketched the requirements for transforming the stream of analog data coming from long-range two-dimensional radar and shorter-range three-dimension radar so that digital data could be sent to the computer.

The heart of the matter with NTDS was, so far as possible, for a battle fleet to continually share data on up to a thousand incoming aircraft or ships. The model pretty clearly was the swarms of Japanese fighter planes that attacked in the Pacific. Any single aircraft could be tracked by two or more NTDS-equipped warships. The job was broadly similar to the SAGE sites keeping watch for Soviet bombers coming over the North Pole, except that with NTDS everything—radar sites, warships, incoming aircraft, and reams of data—was quite literally in motion. The computer calculations about where to aim antiaircraft gunnery or where to send defending planes, for instance, needed to account for the fact that the radar sites on the warships were themselves moving. McNally and Svendsen's fifty-page report, approved by the chief of naval operations in April 1956, required just one alteration before it was given a green light. In the nuclear age, the navy wished to extend the operating distances for fleet activities from the twenty miles that was characteristic during World War II up to 150 miles to more readily cover the vast expanses of the Pacific Ocean. With the selection of Univac in St. Paul as the principal contractor in December 1955, NTDS was set in motion. The navy's officers, located in St. Paul at the ERA–Univac factory, were well versed on the company's unusual technical capabilities. ERA–Univac was then working on the transistor-based Athena missile-guidance computer, whose "outstanding feature is reliability."[30] No other computer manufacturer at the time had a credible plan for achieving such stringent operations requirements with transistors.

Seymour Cray's role in NTDS, while substantial, needs a careful examination. Many well-placed commentators have given him far too much credit. The military-computer expert Ed Thelen probably overreaches with his claim that for NTDS "Seymour Cray was the primary logic and circuit designer." Famed computer designer Gordon Bell suggests that the "Univac NTDS [was] Cray's first computer," which is strictly true only if you give full credit to the prototype, although Cray himself identified the earlier ERA 1103 as his first.[31] What is clear is that sometime in 1956 Cray made a judicious choice to use germanium-based surface-barrier transistors and that he squeezed the utmost available performance from them. He quickly grasped the navy's concept of a "unit" computer and devised a logical structure and architecture that facilitated an entirely novel form of computer networking. While computer "networking" at

FIGURE 3.7. Univac Athena missile-guidance computer, 1961. Athena was a special-purpose missile-guidance computer with eight hundred transistors and four thousand diodes, designed to be an "on-line machine that runs synchronized with the guidance system." Univac's highly reliable design for Athena aided in the complete redesign of Cray's NTDS prototype. Source: Weik, *A Third Survey of Domestic Electronic Digital Computing Systems*, 60.

the time typically meant the use of standard telephone lines, such as those used in SAGE, the NTDS prototype was designed to permit the direct memory-to-memory exchange of data. The then-current generation of germanium transistors allowed a maximum word length of thirty bits, an odd count that nonetheless persisted in the next *six* generations of Univac's military computers.[32]

There is no way to appreciate Cray's genius but to dive into the details. Cray significantly modified the navy's original specification for a split-speed memory, where, to save money, only around one-sixth of the total magnetic-core memory would be expensive fast access while the rest of the memory would be cheaper slow access.

Cray understood that the two memories' different access times would bedevil programming of the machine and that affordable high-speed core memories would soon be readily available. He also suggested rounding up the memory size to the "even" number of 32,768 words to simplify addressing and programming; this number can be represented with fifteen binary digits, and so could each of the computer's memory locations. Some latter-day commentators even point to the unit computer's relatively small number of sixty-two single-address instructions as an early instance of "reduced" instruction set computing, or RISC, later made famous in the mid-1970s by IBM and other manufacturers.[33]

Cray's true genius was not only in squeezing maximum performance from the current generation of transistors, themselves still something of an experimental technology, but also in devising an architecture that permitted two very different types of networking. Even if the unit computer was the "brain" of NTDS, it needed still to be connected by "nerves" to the radar and radio units physically located around the ship; and so the computer's output channels required circuitry capable of sending binary electrical signals down three hundred feet of shipboard cables. Cray opted for a diversity of data widths; high-capacity radar signals could come in on thirty-bit wide channels while keyboards, paper tape units, and other peripherals could rely on narrower six-bit and fifteen-bit channels. The direct memory-to-memory networking was a real departure from anything commonly done at the time. In the case of the landlocked SAGE computers, a duplexed design with two parallel computers was necessary because as late as 1961 the "average error-free running period [was only] 13 hours."[34] When one SAGE computer crashed, the second one took over; but they did not work in tandem.[35] The NTDS "unit" computers were designed from the start to work in tandem. The performance of the "unit" computer was set by the original navy specification that anticipated the smallest installation (aboard a guided-missile frigate) would have two unit computers. Larger ships with more demanding computing requirements would simply have three or more unit computers to achieve the needed computing power.

The first four of Cray's prototypes, configured something like a four-foot tall chest freezer, featured an ungainly "dipstick" for reaching deep into the machine to measure the tiny signals passing between the circuit components. The final two copies of Cray's prototype switched to a space-saving upright-freezer shape with vertically opening access doors. The idea was to conserve scarce floor space on ships. With the new configuration, the computer's circuit boards mounted on horizontal trays could be quickly swapped in and out. Physically, at least, these last two copies of Cray's prototype are not easy to distinguish from the later production versions. With his creative work mostly complete, Cray tried to leave the Univac–NTDS effort in September 1957, shortly after Control Data was organized (see chapter 4), but his actual departure for Control Data was delayed for some months.[36]

86 CORPORATE COMPUTING

FIGURE 3.8. NTDS test facility in San Diego, circa 1960. A shore-based mock-up of NTDS was installed in San Diego for testing and training. Source: U.S. Navy.

A noteworthy departure from Cray's original design came two years later in 1959 after a group of navy engineers took a hard look at the prototype's performance at the San Diego test facility. The test results were not pretty. The prototype did not have a sufficient number of the high-capacity thirty-bit input–output data channels, and the additional narrow channels were simply not needed. The germanium transistors seemed to have hit a performance plateau in power and reliability. And no one liked the awkward "dipstick." In late October, the navy's NTDS project office made a risky decision to redesign the entire unit computer around a brand-new silicon-based "planar" transistor, recently invented by Fairchild Semiconductor's Jean Hoerni. Nearly everything in the computer's design would be up for grabs—except the fundamental thirty-bit word length and instruction set. By December the navy wrote up a revised contract for Univac to design, build, and deliver sixteen of the new silicon transistor computers. The new computers were needed in short order because three test ships were ready for them to be installed, not to mention the shore-based testing, training, and programming sites.[37] The only sticking point was that the "new" design was still on the drawing board.

FIGURE 3.9. Univac's transistor-based NTDS unit computer was equal in performance to the oversize vacuum-tube SAGE computer—at 1/1000 the size, weight, and power requirements. Source: Weik, *A Third Survey of Domestic Electronic Digital Computing Systems*, 56.

The first six months of 1960 in St. Paul echoed the all-out engineering effort, exactly ten years earlier, that launched ERA's pioneering Atlas computer. The navy accepted a preliminary design drawn up by Vernon Leas, who was pulled from Univac's Athena missile-guidance program to work on NTDS. Arnold Hendrickson became NTDS engineering director, and set two dozen engineers to work on the electronic and mechanical design work. He physically moved the three engineers responsible for the core electronic design into the same room to ensure that they would work closely together, while the memory designer got a room of his own. The circuit team devised forty-nine types of printed circuit cards; each unit computer was comprised of 3,810 such cards—secure homes for the ten thousand transistors, thirty-three thousand diodes, and other components—to withstand the pitching and tossing of the shipboard environment. By June, when there were parts enough to build the first prototype of the new design, the four computer designers "inserted most of the cards themselves, one drawer at a time," testing their creation tray by tray as they went. When they were done, the redesigned machine's actual speed of one hundred thousand instructions per second was twice as fast as Cray's prototype and also, remarkably enough, significantly faster than the enormous SAGE computers that were at the time lumbering off IBM's assembly lines.[38]

In October, with the engineering work winding down and attention shifting to manufacturing, Univac submitted a fifty-eight-page project report to the Bureau of Ships. It detailed the basic building block for the navy's tactical combat system or NTDS. The entire computing package was squeezed into a compact upright cabinet thirty-three inches deep and thirty-seven inches wide, or a bit more than eight square feet, and it consumed just 2,400 watts—around one thousandth the size and power-requirements of SAGE. Thirteen horizontal trays housed the logic and ferrite-core memory modules, with test points easily accessible at the front that put the troublesome dipstick out to pasture. Networking with the external radio, radar, or threat-tracking equipment was handled by twelve input and twelve output data channels, capable of maximum transfer rates up to thirty thousand words per second, with everything standardized at thirty-bit words and channel widths. (Four additional channels were available to directly link one unit computer with another.) The computer's control program paused momentarily to set up a data transfer, but then returned to program execution while the data streamed into the computer's internal memory or external storage. Notably, "external equipment [had] direct access to the computer's internal memory" owing to a priority-routing scheme that did away with input–output data buffers. An internal seven-day clock could initiate routine programs or even log the arrival of an incoming data stream; obviously, such a seven-day feature presumed that the computer was capable of extended round-the-clock operation. And for real-time tracking and targeting of incoming aircraft, the computer itself had a full complement of trigonometric functions as well as conversions between circular and square coordinate schemes.[39] Manufacturing and delivery of the sixteen NTDS computers, despite some bumpy patches, was complete by June 1961.[40]

Expansion of Univac's computing projects in the 1960s led to its spreading out beyond its urban facilities in St. Paul. In 1961 the Univac data-processing division moved into the newly constructed Plant 4 in north suburban Roseville with five hundred employees (see chapter 4 for Control Data's move to the suburbs the same year). Six years later, the tagline was "The Story Is Growth" because three thousand employees—assemblers, technicians, clerks, engineers, and administrators—occupied 490,000 square feet in "two attractive buildings" on Highcrest Drive looking over a golf course. Univac shared the Roseville industrial park with regional technology heavyweights Honeywell and 3M. In 1967, Univac's defense systems division moved into a brand-new $3.5 million "Univac Park" complex in south suburban Eagan. (The I35-E freeway was under construction at the time, too.) A sign of the time was that parking for 1,200 cars—one car per employee—was available just off Pilot Knob Road. Whereas the Roseville facility was for factory production, the Eagan site was designed for engineering, office, and laboratory space. By 1968, Univac, spending roughly $120 million each year on payroll, purchases, and taxes, employed a total of 10,500 people in the Twin Cities area, which was the firm's largest concentration of workers "in the world."[41]

NTDS Programming and the Real World

It is worth recalling that software for any sort of networked real-time system such as SAGE or NTDS had never been conceived or built before. Computation at the time was mostly in a "batch" mode where financial computations, data sorting and manipulations, or scientific calculations were completed in sequential order. If there was an error in the program, it could simply be corrected and redone. (Early experiments at MIT and Dartmouth in time-sharing computers were under way, as chapter 7 recounts.) Obviously, real-time battle management on land or sea was entirely different. While timely coordination of incoming radar signals from enemy bombers for SAGE was a significant challenge, with NTDS there was the additional complication that the radar installations themselves were on moving warships. "The programming difficulties in 'tracking' the radar contacts, or determining their expected position by computer calculation so that each radar blip could be associated with an already identified ship or plane, proved to be much more difficult than anticipated," one participant recalled.[42]

State-of-the-art programming at the time entailed "compiling" human-readable instructions into machine-executable binary code. NTDS featured three levels of compliers including a "high-level language [CS-1] which allows programs to be written by means of powerful operators totally divorced from machine code."[43] Such operators could create a computer program with logical structures such as if/then, go-to, switch,

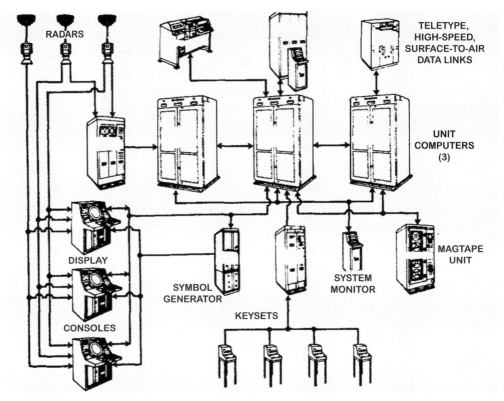

FIGURE 3.10. Univac NTDS system configuration, 1961. The NTDS computer was designed "to remove from the [human] operator, to the maximum practicable extent, tiring and repetitive operations in order to concentrate his effort in areas requiring . . . judgment and experience." Source: George G. Chapin, "Organizing and Programming a Shipboard Real-Time Computer System," *Fall Joint Computer Conference Proceedings* (1963): 127–37, on 128.

or return. Data tables, something like an early database, could be created, searched, modified, and deleted. And special built-in routines handled errors in division or multiplication, adding a software component to the system's distinctive reliability. NTDS programmers constructed well-defined operational "modules" or subprograms that communicated in carefully defined ways with the "executive" or operating system.

An NTDS computer might at any time be running between ten and twenty-five subprograms (such as navigation, automatic tracking, or intercept control), with the executive program allocating actual computational times based on subprogram priorities and existing wait times. For instance, during periods of heavy computational load, external displays might be updated at a slower rate than the standard twenty times per second.[44] For the most part, extensive Univac–navy cooperation ensured that the NTDS "unit" computer communicated flawlessly with external equipment such as the radar units made by Hughes Aircraft and the intership communications equipment made by Collins Radio. A last-minute "fix" was needed when Univac engineers discovered that a Hazeltine display console used an incomprehensible mixture of straight binary data and one's complement notation to represent x- and y-coordinates and velocities.[45]

While SAGE never faced down a fleet of incoming Soviet bombers, the Vietnam War provided a severe real-world test for NTDS. Ten navy ships were already equipped with NTDS by December 1965, with an additional fifty-five installations either under way or planned. NTDS composed the navy's tactical brains for the duration of the conflict.[46] The coastline of North and South Vietnam was split into three zones, with all of the NTDS ships linked by a secure high-speed communications network. The central Positive Identification Radar Advisory Zone (PIRAZ) was overseen by one of five ships including the USS *Long Beach*, where, in the NTDS-equipped Combat Information Center, as one navy officer recalled, "It was uncanny to see, in real time, what was going on over a war zone several hundred miles long. You could even see the tracks of the 16 inch shells that the battleships were lobbing over Vietnam." At the peak of the war, thrice-daily bombing raids over North Vietnam involved hundreds of airplanes. The PIRAZ ship simultaneously tracked air force planes coming from land bases in Thailand and Laos, navy planes flying from aircraft carriers off South Vietnam, South Vietnam's own military aircraft, and various commercial airplanes. Enemy MIG fighters were targeted for missile attack by special long-distance radar. Data from other ships in the region, or an airborne air force reconnaissance plane, often assisted with dispatching search-and-rescue helicopters to rescue downed pilots. "Having the NTDS system down would have taken our ship out of the action completely, however I do not ever remember this happening during two nine-month deployments."[47]

Because NTDS evolved into a specialized air-traffic control system, it was not terribly difficult for Univac to adapt its accumulated expertise for civilian air-traffic control. Indeed, with a hope of solidifying congressional funding for NTDS, the navy pointedly encouraged Univac to actively bid for civilian air-traffic control. Already by 1960, after securing the contract for the Federal Aviation Agency (today's FAA) phase-one in-route system, Univac File computers at five airports were tracking airplanes between cities, with an experimental installation also at Atlantic City, as well as maintaining ticket reservations for three major airlines. With the superreliable NTDS under its belt, Univac was well positioned when the FAA announced plans for a national air-traffic control system.[48] While IBM secured the main contract for the in-route (between cities) system, Univac secured the main contract for the Automated Radar Tracking System (ARTS) that controlled landings and takeoffs at each individual airport; it installed an NTDS-derived computer for the first experimental system in Atlanta in 1964. In 1969, Univac won the follow-on $35-million contract, subsequently installing ARTS-3 in sixty-three additional cities in the next six years. For ARTS-3,

92 CORPORATE COMPUTING

Sperry Rand replaced the NTDS-vintage computer with an updated Univac 8300 computer, while retaining its proprietary Ultra assembly language for the complex software.[49]

Despite impressive leaps in computer hardware during these years, it proved excruciatingly difficult to subsequently revamp the Air Traffic Control (ATC) software for either the in-route or terminal approach systems. In effect, the ATC system became locked in to 1960s software. In the mid-1990s, the FAA was forced to cancel its multibillion-dollar Advanced Automation System, more than a dozen years after it was started, when it could not overcome crippling cost overruns and unsatisfactory performance issues.[50] Consequently, even with the looming threat of the Y2K crisis, because ARTS-3 was still landing planes at fifty-five airports, the FAA was still purchasing Univac 8300s as late as 1997. Remarkably enough, thirty-bit Univac terminals,

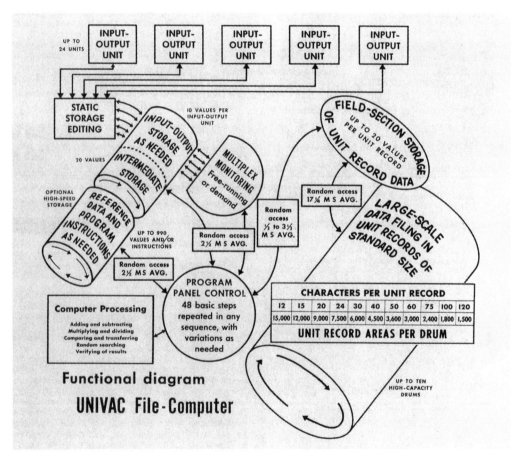

FIGURE 3.11. Univac File Computer (1961) was a "general purpose, medium-priced electronic data processing system with a magnetic drum memory." Univac used the File computer for early air-traffic control. Source: Weik, *A Third Survey of Domestic Electronic Digital Computing Systems*, 932, 934.

directly descended from Cray's thirty-bit NTDS prototype, remained in service.[51] Japan, Korea, Taiwan, Germany, and China also installed ARTS-related systems. There is even, well after the Y2K crisis, one unconfirmed report of an original NTDS-vintage computer landing airplanes in Oklahoma.[52]

Quality Revolution

When the NTDS computer went onto the warships and out into the field, one of the striking and unusual findings was its rock-solid reliability. Already by the early 1960s, its typical measure of average "mean time between failures," or MTBF, was surpassing an astonishing one thousand hours. The first seventeen NTDS computers in the field ran reliably between 1,500 and 2,500 hours each, experiencing a failure just once in every two or three *months* of continuous operation. The extraordinary reliability gave rise to some odd problems, because in the first instance naval technicians had a difficult time gaining experience in troubleshooting the NTDS computers. The navy chain of command became suspicious because, while they received regular reports on the failures of radar scopes and computer displays, they were not receiving any reports on NTDS computer failures. When grilled by his commanding officer, one ship's chief electronics technician explained, "Hell, it don't hardly ever fail, sir."[53] These unusual failure rates called for official comment. "This high figure is attained through complete analysis of every component that fails anywhere in the world," observed one government analyst. "This information is transmitted to venders for corrective action when necessary." Noting the "exceptional" reliability of the NTDS equipment "at apparently no excessive cost," the navy recommended that "reliability methods used by this contractor should be explored for use by others."[54] Univac's aggressive quality control helped bring about a transformation in the entire semiconductor industry. It started with the painstaking work of the NTDS Reliability and Failure Analysis Group.

This small group, initially just three engineers and a technician, sought compelling visual evidence to force individual semiconductor vendors to address the evident flaws and inconsistent quality of their products. It is fashionable to laud the industry's fabled progress in making components smaller and smaller—now canonized as Moore's Law—but it is doubtful that any such progress could have occurred absent a massive upgrade in quality production. Typically, a vendor such as Motorola or Sylvania aimed to ship as many diodes or transistors as possible, knowing full well that some of them were of marginal quality and would be rejected by the computer manufacturer well before ever seeing service. Univac's quality engineers flatly rejected the industry's commonplace notion of "random failure," striving to identify and trace the root cause of each and every device failure. They went to work with jeweler's tools, dental X-rays, and microphotography. After painstakingly slicing open the metal caps on thousands of individual transistors, they found inside foreign

94 CORPORATE COMPUTING

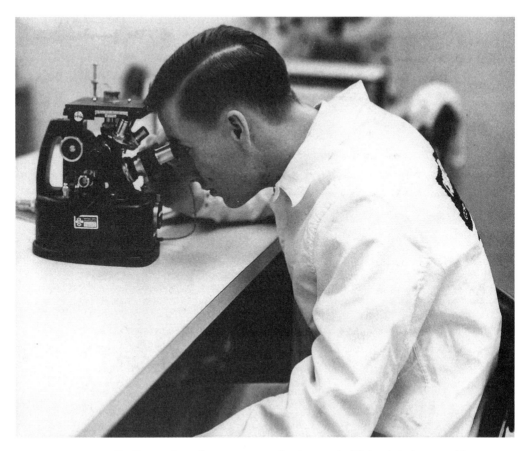

FIGURE 3.12. Quality testing of government circuitry, 1964. Univac's industry-wide quality efforts in the 1960s paved the way for the semiconductor industry's later advancements known as "Moore's law."

junk—dirt, metal bits, human hair, insect parts, and worse—that did not make for polite dinner-table conversation. Vendors might reasonably argue that a certain number of transistors were statistically bound to be defective, but it was difficult for them to dismiss the gruesome photographic evidence.

Over time, the semiconductor industry instituted closer control of manufacturing conditions, leading eventually to the obsessively spotless "clean rooms" that, as one author puts it, are "essential to high tech manufacturing."[55] Another, more subtle problem addressed through purposeful interaction between semiconductor vendors and computer manufacturers was the affliction colorfully known as "purple plague." With silicon transistors becoming ever-more important in many applications, it was especially worrisome that purple plague occurred only in silicon transistors. At an American Society for Quality Control conference in 1962, Univac engineers pointed to an unusual chemical reaction that occurred when gold wires were bonded to aluminum contacts in silicon-based transistors. Such a device might behave perfectly

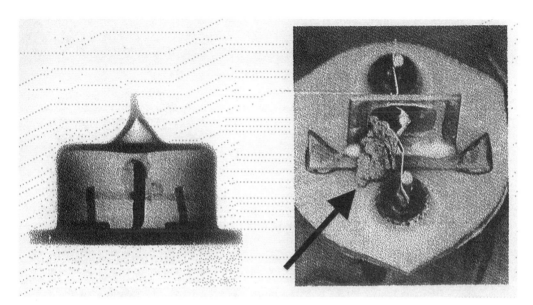

FIGURE 3.13. Univac photography of "junk" in transistors, 1962. "The X-ray of Figure 29 [at left] shows the presence of a foreign object inside the shell of a transistor [that failed in service]. Figure 30 [at right] shows the opened transistor with . . . an amorphous chip of stainless steel. Because no stainless steel is used in manufacturing the device, the supplier was at a loss to explain its presence." Source: George P. Anderson, "Failure Modes in High Reliability Components," *Transactions of the American Society for Quality Control 1962* (Cincinnati): 353–62, on 359.

well when it first came from the production line but then deteriorate markedly after only a few months of service. For a company whose NTDS computers ran for months without any failures at all, the purple plague was especially troubling. Ultimately, research by Univac, Sylvania, and others convinced the semiconductor industry to banish the use of gold wires inside silicon devices and entirely avoid the troublesome gold–aluminum junctions.[56]

Having secured its own reputation for government-sector computing, Univac also helped bring about a "quality revolution" in the nation's semiconductor industry. Created in 1972 to centralize the purchasing of literally millions of discrete diodes and transistors for the far-flung Sperry Rand factories, the Semiconductor Control Facility (SCF) located in suburban Roseville, Minnesota, soon became a means for standardizing the huge variety of components routinely ordered by the company as well as hounding semiconductor vendors to improve device reliability and performance. Staffed eventually by two hundred of the company's quality-control experts, the SCF looked both inward and outward. Looking inward, it saw that Sperry was using a very large number of nonstandard components and so paid higher prices for their delivery. It sought to form standardized "lists" of the highest-volume components, both military and civilian, and to negotiate bottom-drawer prices on them

FIGURE 3.14. Disk file storage unit assembly in clean room, circa 1962. High-tech manufacturing gained a reputation for being especially "clean" with such facilities.

from the national vendors. The vendors saw an advantage in that they might be able to keep an entire production line busy with just a single standard component, increasing their output of quality components, rather than shift a production line between multiple components. SCF impressively argued that manufacturing and purchasing costs would go down and that component quality would improve; in exactly these terms it amply justified its hefty budget (around 10 percent of total semiconductor purchases) for fourteen years. It also closely monitored the quality of each and every incoming batch of semiconductor components.

Looking outward, the Semiconductor Control Facility made regular reports to each of the vendors—all the big national names, including Texas Instruments, Motorola, Fairchild Semiconductor, and others—comparing each company's reject rates on every major class of component to the other companies'. An SCF report to an individual vendor listed the name of only that company with the other companies' names blacked out, but the internal records kept by Sperry Rand reveal all the gory details. The SCF's successful behind-the-scene efforts, oddly enough, were

Plant Number	Location	Employment
Plant 1	2750 West Seventh Street, St. Paul	2,100
Plant 2 (original ERA)	1902 West Minnehaha Avenue, St. Paul	500–749
Plant 4	2276 Highcrest Drive, Roseville	3,400
Plant 5	640 North Prior Avenue, St. Paul	500–749
Univac Park	3333 Pilot Knob Road, Eagan	1,100

FIGURE 3.15. Univac plants, locations, and employment in Minnesota, 1971. Source: Univac documents; *Minnesota Directory of Manufacturers* (1972).

not publicly acknowledged. Univac's industry-shaping quality efforts, owing to their sensitive nature and proprietary consequences, were never widely discussed in the professional literature or trade press, according to longtime SCF manager Mike Svendsen.[57] Perhaps the most durable innovation in the computer industry—imposing an industry-wide mechanism for making objective analytic assessments of quality for all the major semiconductor vendors—could never be properly acknowledged.

4
Innovation Machine

Control Data's Supercomputers, Services, and Social Vision

A calendar from 1957 brims with events that reshaped the landscape of computing in Minnesota and the world. The events rippled outward from the Soviet Union's launch of Sputnik. The tiny radio beeps that Sputnik sent down seemed ominous to many Americans, who reasoned that if the Soviets launched satellites up into space it wouldn't be too long until they might aim missiles down on North America. A new phase of the Cold War was at hand. Immediately, the U.S. Department of Defense created its Advanced Research Projects Agency with a mission of conducting farseeing military research, and soon the country's aeronautics agency was transformed into a full-blown National Aeronautics and Space Administration. While NASA directly funded the space race and created a huge market for computing and controls, ARPA literally created the field of computer science with cutting-edge projects in computer time-sharing, graphics, and artificial intelligence, as well as computer networking that eventually became the Internet.

Given this fertile environment of lavish government funding, urgent national missions, and rapidly growing markets, it is no accident that three remarkable U.S. computer companies were founded in 1957. Together they would reshape the nation's and the world's computing. Digital Equipment Corporation was the product of Boston venture capital and MIT entrepreneurship. DEC, as it was commonly known, dominated the field of minicomputers with its PDP and VAX series of machines, and it was for many years the anchor firm of metropolitan Boston's "Route 128" high-technology district. In California, Fairchild Semiconductor was a spin-off from the research laboratory founded by William Schockley, the famed coinventor of the transistor. In addition to inventing the silicon transistor that found its way into NTDS (chapter 3), Fairchild built the first commercially successful silicon-based integrated circuit and was the platform for Robert Noyce and Gordon Moore launching a second-round spin-off in 1968 initially called Integrated Electronics Corporation. Better known as Intel, it did much to create the field of integrated circuit memory; and the firm, one of the anchors for Silicon Valley, subsequently dominated the world market for microprocessors.

Control Data Corporation, formed in downtown Minneapolis and staffed largely by managers and engineers from ERA–Univac in St. Paul, was the third notable computer start-up of 1957. The timing was propitious. The very same issue of *Electronics* that announced its formation was packed with the technological opportunities of the Cold War. It was a high-tech roundup that promised a boundless future for the new company: ICBM missiles and missile-defense systems, Sputnik satellite tracking, an "atomic bonanza" in Europe, forecasts of military research spending in communication and data processing, and direct coverage of how "pressure from Congress is expected to hurry U.S. space satellites."[1] This chapter examines Control Data's thirty-year history as an innovation machine.

Lives and Legends

The most frequently told origin story about Control Data goes something like the following, and it's mostly true. Famous companies need famous founders and Control Data had two: William C. Norris and Seymour R. Cray. Cray and Norris were dissatisfied with Sperry Rand's inept management and timid strategy, and they founded Control Data Corporation in 1957 as a way to re-create something of the glory days of ERA. Their motivating idea was that technical experts who knew where technology was headed should be the ones directing high-technology companies. With Cray heading the company's engineering and design, and Norris providing strong executive leadership, Control Data brought to market an amazing string of computing machines. The first was the all-transistor CDC 1604, named in another mathematical pun—or so the apocryphal story goes[2]—when the company's new address of 501 Park Avenue in downtown Minneapolis was added to the ERA–Univac 1103 model. Fame and fortune came with the superfast CDC 6600 and the successor CDC 7600, widely hailed as the pioneering "supercomputers." Massive profits from these premium-priced machines made Control Data into a stock-market darling, with the firm surpassing $1 billion in annual revenues by 1969.

Much, but not all, of this origin story is true. Norris and Cray were certainly the firm's dominant figures and public faces, but no two men can make a billion dollars appear. Cray was still finalizing his work on the NTDS prototype computer, as chapter 3 related, when Control Data was founded in the summer of 1957. Norris and the managers at Univac each hoped they would gain or retain Cray's unique talents—even if he seemed destined to follow Frank Mullaney and the other ERA spirits to Control Data. One estimate is that "Cray arrived in late 1957," perhaps four or five

FIGURE 4.1. Control Data's debut as "CDA" on the New York Stock Exchange, March 6, 1963. From left: Edward C. Gray, NYSE vice president; William C. Norris; Robert J. Silver, specialist trader.

months after the company's founding.[3] Even Norris himself, who was Control Data's president from 1957 to 1986, did not sign the incorporation papers. The impressive 1969 annual revenue of $1 billion cannot possibly be credited solely to the supercomputer business. You simply do the math. CDC sold around one hundred copies of the CDC 6600, beginning with its first sale to Lawrence Livermore in 1964, at about $7 million each, for estimated revenues of $700 million—spread across five or six years. (The successor CDC 7600 first shipped in 1969.) To fully understand Control Data we will need to look beyond its fame to its broader range of activities.

It is insightful to conceptualize Control Data's creation as a product of the "push" of corporate politics at Sperry Rand with the "pull" of a supportive business and financial environment in the Twin Cities, with Bill Norris at the center of both. What became Control Data took protean form early in 1957, when two ERA veterans, Willis K. "Bill" Drake and Arnold "Bud" Ryden, met for lunch in New York and talked over the worrisome situation. They knew that Norris was battling another round of reorganization at Sperry Rand, and the rumor mill had it that massive resignations in St. Paul were imminent.⁴ Even though Drake had left St. Paul years earlier to work at General Electric's pioneering Univac installation in Louisville, Kentucky, he had remained close to the St. Paul engineers, including Norris. Ryden, a Harvard-educated MBA and former ERA treasurer, had been negotiating government contracts for Honeywell and was doing some part-time consulting for Norris. Over lunch he listened carefully to Drake, then boldly broached the idea of a new company. As Drake recalled, "he planted a seed on that day."⁵ Ryden and Drake talked informally with Norris, who was still trying to make things work with Sperry Rand, as well as with Frank Mullaney and others. Everyone liked the idea of a "new company to be a vehicle for the talent group that had worked together in Univac and earlier days in ERA," but no one could figure out a plan for raising six hundred thousand dollars to launch the new venture. Pretty clearly, Univac would see the new company as poaching its employees and bring legal action, chilling the already uncertain prospects for fund-raising. In mid-April, Norris lost the reorganization battle, effectively getting fired from his position as Univac vice president and division manager. The ERA engineers saw it that "Norris' job was being cut way back, and that he was being sent back to the boonies." The die was cast: "when [Sperry consultant] Thornton Fry came again with sledgehammers, the formation of CDC was the result."⁶

Ryden and Drake, setting aside their other duties, devoted full time to organizing the new company. On the fourteenth floor of the Minneapolis Athletic Club (today the Grand Hotel) they juggled a list of "investor sizzle words," before deciding on "Control Data Corporation."⁷ Then, on Drake's kitchen table they drafted an unusual stock prospectus. They found a little-known state law that permitted stock sales solely to Minnesota residents, conveniently enough, with no public filing to the Securities and Exchange Commission. With any luck, Univac would not find out about the new company until it was too late. It was a risk to employ this legal gambit, but several points were in their favor. The memories of early ERA investors getting a 3,000 percent return certainly loosened purse strings and wallets, and with stock priced at one dollar per share there was plenty of room for small investors. Ryden even arranged with a friendly vice president at the First National Bank of Minneapolis so that fifteen engineers could each purchase five thousand dollars in stock by putting down only one thousand and borrowing the remainder. Drake went door-to-door selling stock, then filled up the subscription book with three evening meetings at his house in Edina. "So we had one prospectus in a ring binder and we followed what was called a pre-registration subscription . . . at that time you could have people read the prospectus and sign the pre-registration subscription without filing anything with the state of Minnesota other than your corporation papers."⁸ The subscription book also filled up with 20,000 shares to partners at the Piper Jaffray investment house, Ryden took 20,000, an Iowa surgeon took 25,000, and Norris himself signed

FIGURE 4.2. McGill Building at 501 Park Avenue in Minneapolis (shown here circa 1962) was Control Data's first headquarters. The original "Control Data Corporation" sign to the left of the entrance door is now at CBI.

up for 70,000, for a total of around 300 investors large and small. Ryden noted, "I wanted to get people with money invested initially and they'd buy $10,000–15,000–20,000, but as it became apparent that the prospective employees were putting in so much, I actually went back to several of those [larger investors] and cut them in half." Even Norris trimmed back to 58,500 shares.[9]

Formally, at least, Control Data was created when three Minneapolis businessmen—Fremont Fletcher, Abbot L. Fletcher, and D. P. Wassenberg—filed the incorporation papers on July 8, 1957. They publicly announced the hiring of Norris as president, five weeks later, on August 14. (During the interval, Norris had resigned from Univac, and wild rumors swirled about his future plans.) In the interim, Fremont Fletcher served as president, and he later continued as secretary and board member. True to form, Sperry Rand was upset when it found that its top talent in St. Paul was resigning in droves, going "over the river" to Minneapolis. "Once we had the money it was microseconds before we got hit with a lawsuit from Sperry charging all kinds of terrible things," recalled Drake. "It was clearly thought of as harassment and an effort to squash this thing before it gets going." But with the six hundred thousand dollar stock issue already in place, there was no way Sperry could halt the new firm outright. Sperry Rand's lawsuits against the new concern charged theft of intellectual property, even though its legal action, like most lawsuits of parent firms against their spin-offs, went nowhere.[10]

Businesses of Computing

In September 1957, the new company with its dozen or so employees moved into an otherwise empty warehouse at 501 Park Avenue in downtown Minneapolis, setting up gray desks and drawing boards and plywood partitions in the shadow of massive rolls of newsprint destined for the *Minneapolis Tribune*. "At night I could hear the rolls slipping every once and a while and there was a block of wood under the end roll and so my fear was that the block would slip some night and all those rolls would come down on me and I be squashed against the far wall," remembered Cray.[11] The July prospectus gestured rather grandly to "the design, development, manufacture and sale of systems, equipment and components used in electronic data processing and automatic control for industrial, scientific and military uses," but it is worth noting that the word *computer* was notably absent—from each of the 1957 start-ups. One of the financial supporters of Digital *Equipment* Corporation actually vetoed the word *computer*, while Fairchild *Semiconductor* described just what that new firm aimed at. For its part, Control Data lacked two crucial items for a successful business: it had no detailed business plan specifying the steps to achieve its lofty goals; and, besides the two thousand square feet of low-rent office space, it lacked facilities to attract business from industry, science, or the military. As one befuddled Minnesota state securities inspector reportedly blurted out, "you don't have any products, you don't have any plant, you don't have any machines, you don't have any customers, you don't have any money. And you want to sell stock in—in what?" The thinking was "we've got to have some credibility . . . be making and shipping something; and we got to do it before we've gone broke," Drake recalled.[12]

In November, possibly even before Cray came on board, Control Data bought up Cedar Engineering, a sizable local machine shop that made motors, servos, and actuators for the aircraft industry. Cedar Engineering, located in the near-west suburb of St. Louis Park, named for the nearby Minneapolis and St. Louis Railway, was itself a five-year-old spin-off from Minneapolis–Honeywell (chapter 5) that added manufacturing and engineering prowess to the metro area's expanding industrial district. Its owner, E. J. "Jim" Manning, a "shirtsleeves tough guy," had gotten into a financial scrape and was looking for an out. The new factory brought its existing revenues, receivables, 165 employees, and fully appointed design and manufacturing space. The haul included a number of notable future Control Data executives, including

FIGURE 4.3. Control Data's machine shop at Cedar Engineering, St. Louis Park, Minnesota, 1959. Cedar Engineering workers exemplified strong mechanical skills developed in the Twin Cities' industrial district. This factory helped Control Data to secure early bank financing and its first military contracts.

Tom Kamp and Ed Strickland, as well as "some very clever Swiss technicians." "When people would come to town they would be taken out to Cedar [Engineering] to see the machinery, and then they would be taken downtown to talk to engineering, and that was a big factor in getting government contracts."[13] Over the years, Cedar Engineering became "a cash cow for the corporation" and a "springboard . . . in the manufacturing area."[14]

Control Data's first year was touch-and-go. Norris successfully signed up Frank Mullaney from Univac, and Mullaney brought over many of Univac's best and brightest, including such future leaders of the company as William Keye, Robert Kisch, James Miles, Henry Forrest, and Seymour Cray. "I came over to Control Data in desperation because things at Univac had become so chaotic from an engineering standpoint . . . everybody had their troubles," recalled one Univac veteran. Another, James Harris, later head of Control Data's Minneapolis data center, recalled: "I got bloody knuckles from pounding on the door trying to get in." Once inside, there was not so very much to see. "We set up a little lab. We came in at night and we put together the benches because they were cheaper that way and so forth. We didn't even have a quarter inch drill," remembered Mullaney.[15]

FIGURE 4.4. In 1957 Frank Mullaney and Bill Norris were guiding forces for Control Data, carrying forward the ERA–Univac ethos.

An early sign of financial stress was Norris's announcement that he was slashing his paycheck and everyone else's to conserve the new company's rapidly dwindling operating funds. By March 1958—even as it announced two hundred thousand dollars in new missile contracts, expanded its employment to two hundred people, and its stock locally traded at a hefty advance of four dollars a share—Control Data was in serious financial trouble. On March 23, Norris went home to St. Paul and typed out a highly unusual two-page letter, most likely to keep it out of the official company records. It was to Henry Forrest, Control Data's contract man in Washington. Owing to the "unanticipated delay" in landing a big computer contract, Norris was alarmed about the company's cash flow. Monthly payroll and expenses had been running thirty thousand dollars, until the recently imposed "drastic curtailment" of all expenses. Even so, it was only three months or less before the company hit a wall. Just two or three weeks of funding remained for Cray's developmental work. Norris noted that "as President of a successful public corporation I can't gamble" on running out of cash. "You will of course treat this matter with absolute secrecy," Norris warned Forrest. "This has a much higher classification than any contract we are about to get. If anything leaked out on this plan the rumors would fly thick and fast." Norris was hoping for a "piece of firm paper," the timing of which Forrest, in the midst of contract-rich Washington, was "a better judge than anyone else." Two weeks later, just before it was hit by the lawsuits from Sperry Rand, Control Data secured some "firm paper" with a one hundred thousand dollar loan arranged by Bud Ryden from the First National Bank of Minneapolis (a bank vice president's visit to Cedar Engineering had worked miracles). The loan provided some breathing room until that summer when several sizable military research and missile contracts came through. Norris then announced his firm's hiring of William G. Shepherd, chair of the University of Minnesota's electrical engineering department, to serve as associate director of research and provide "technical counsel on advanced research."[16]

Sometime early in 1958, Cray made an outlandish suggestion. "Look, we don't have any jobs anyway, why don't we build a computer?" as Mullaney recalled it.[17] Cray had been tinkering with transistor circuits that might form a small computer, nicknamed the "Little Character." "The idea was that to a really sophisticated buyer . . . that demonstrator, which we could afford, would be enough in order to warrant actually getting some purchase orders. Around the purchase orders you can finance, and so 'Little Character' was kind of a boot-strap approach," another insider recalled. During these anxious months, there was nothing to sell. Employee number one, Bill Drake, who had made a name for himself in marketing and sales during the Univac days, was let go; and Norris and Ryden to all appearances tangled over the intangibles of foundership. "Norris really wanted to be identified as the sole founder which meant he'd have to get rid of all the guys, and if he got rid of Drake, I'm the next one," Ryden remembered.[18]

FIGURE 4.5. "Little Character" computer exhibit, 1984. Robert M. Price presents a prototype of the earliest CDC computer, built in 1957–58, to Gwen Bell and the Computer Museum of Boston.

Pushed out from the inner circles of Control Data, Drake and Ryden founded the Midwest Technical Development Corporation in October 1958. Joined by CDC founder Fletcher Fremont, they aimed to replicate the pioneering venture capital firm American Research and Development Corporation, or ARDC. ARDC had been created in 1946 by Harvard Business School professor Georges Doriot, often called the "father of venture capitalism," with the assistance of former MIT president Karl Compton and the engineer, banker, and Vermont politico Ralph Flanders. It so happened that ARDC backed the 1957 start-up Digital Equipment Corporation.[19] Ryden, with an MBA degree from Harvard, was well acquainted with these developments and hoped the new venture would "accelerate development of Minnesota's growing roster of small scientific and technical industries." By 1961, Drake noted that MTDC was "one of the most successful young venture capital firms in the country," with more than five thousand shareholders and a portfolio of twenty start-up companies. MTDC's most notable success came after three Ph.D. scientists from Sperry Rand's Norwalk research laboratory sought venture funding from it, heightening the impression, in Connecticut at least, that more Minnesotans were poaching its technical staff. In the event, the Sperry scientists mixed with Minnesota money to form National Semiconductor and, once again, become the target of a Sperry Rand lawsuit.[20] MTDC principals also invested heavily in Telex, a St. Paul manufacturer of hearing aids that they remade into a full-scale electronics and computer component manufacturer. It is massively tempting to speculate what might have happened if MTDC—the second publicly financed venture capital firm in the nation—had not been broken up by a lawsuit in 1963.[21]

Given the fledgling Control Data's unsteady finances, the award of several large government contracts came just in time.[22] In addition to its technical prowess, Control Data also had political experience in contract negotiation and technical liaison in Washington owing to the ERA–Univac veteran Henry Forrest. His well-honed contacts in Washington were "very helpful" in the early years, as Norris had clearly appreciated during the salary crisis of 1958. "On the premise of expanding military

FIGURE 4.6. Control Data 1604 serial 1, 1960. The first CDC 1604 arrived at the Naval Postgraduate School in Monterey, California, in January 1960. Its notable success brought much-needed revenues and national distinction to Control Data.

expenditures . . . military business will be a substantial position of the total business for many years," stated Control Data's early prospectus. "Great care will be exercised in the development of commercial business." In this respect, Control Data mirrored Fairchild Semiconductor's early dependence on military markets. The company's financial prospects brightened measurably with the award of an air force contract to Cedar Engineering in January 1959, followed by a $2.5 million navy contract for the first CDC 1604 computer later that year, delivered to the U.S. Naval Postgraduate School in Monterey, California. Several sizable missile and satellite contracts came in the following year. Government contracting contributed handsomely to the company's bottom line in 1961 with a $5-million contract for fire-control computers for the Polaris submarine and, a few years later, by 1965–66, a $14-million award from the U.S. Atomic Energy Commission and a $22-million contract from the United States Post Office. The deep-pocketed Lawrence Livermore national laboratory bought only Control Data and Cray computers, with the exception of two IBM 7094s, for two decades from 1963 to 1982.[23]

FIGURE 4.7. The first Control Data 6600 supercomputer was shipped to Lawrence Livermore National Laboratory in 1964. Seymour Cray and Jim Thornton are on the loading dock in doorway at left.

In the 1960s, the manufacturing activities at Cedar Engineering evolved into one of the three core businesses that created and sustained Control Data. (The other two lines of business, computer services and supercomputing, are examined later in this chapter.) Cedar Engineering's Tom Kamp moved up through the managerial ranks at Control Data, emerging as the top executive for CDC's expansive Peripheral Products division. It is little remembered today, and it was insufficiently appreciated at the time, but, put simply, Control Data needed a dependable and steady source of revenue in order to finance the pricey rounds of high-end computer development. Years of expensive development effort were needed before each new-generation computer was ready for sale. Even though Cray prided himself on his independence from corporate politics and structure, the entire Control Data enterprise needed a steady stream of cash for such prosaic items as payrolls, rents, equipment, and supplies. "The dollar penalty for mistakes in designs like large computers is absolutely tremendous," recalled one of Cray's engineers.[24] Ultimately, Control Data's innovation machine depended on the cash flow from the sales of peripheral products such as tape drives and memory units.

FIGURE 4.8. Complete disk-drive units roll off assembly line, 1972. High-volume manufacturing permitted Control Data to dramatically boost output and lower unit costs. Mass manufacture of computer peripherals was one of the company's three strategic pillars that turned it into a billion-dollar operation by the late 1960s.

To ramp up its production of peripheral products, Control Data planned a $1.5-million manufacturing facility that was clearly impossible to accommodate at either its downtown building or Cedar Engineering. The effort launched Control Data into the Twin Cities suburbs. Many of America's leading corporations in these years built attractive "suburban campuses" outside the established urban centers; and, seemingly the future, Minnesota's Southdale Mall opened in 1956 as the world's first indoor enclosed mall. But Control Data's move to the suburbs four years later did not go smoothly. More than a dozen sites in the Twin Cities were scrutinized for their location and amenities. Although Control Data hoped to build its new factory in north suburban Roseville, the residential location (just south of Highway 36) provoked worries about the impact of a thousand new workers and an increased load on local roads and schools. A citizen's group blocked the city's efforts to rezone the residential parcel into an industrial site.

Control Data went back to its list of possible sites. It happened that south suburban Bloomington was willing to set up a 250-acre "industrial park," for Control Data along with several smaller manufacturing businesses, that was located east of the Metropolitan Stadium (the site today of Mall of America). In March 1961, when its stock price topped an eye-popping $101 a share, Control Data published plans for the three-story headquarters office building (later becoming the company's world headquarters [Figure 4.19]) and adjoining two-story production facility with one hundred thousand square feet. Governor Elmer Andersen sent Norris a congratulatory letter praising him for putting a thousand Minnesotans onto the company's payroll. "All of Minnesota applauds the success and phenomenal growth of Control Data," he noted. "We are eager . . . to do all possible to aid in the continued expansion of your corporation."[25]

With the ink barely dry on the first suburban facility, Control Data made plans for a second new factory in Bloomington–Edina. At this suburban site, located immediately northeast of the intersection of today's highways 100 and 494, Bloomington created another multitenant industrial park (among the tenants were the University of Minnesota's spin-off Rosemount Engineering) fronting the aptly named Computer Boulevard. There, by the autumn of 1961, Control Data had moved into the seventy-six thousand square-foot facility and was filling so-called OEM agreements.[26] (OEM stands for original equipment manufacturer, but the precise origin of the term is murky.) Control Data physically made these items, but another company, such as NCR or Honeywell, rebranded them as part of a larger system or installation. The first such product announced in 1960 was a model 180 production data recorder developed by Cedar Engineering. An early lucrative line was in magnetic tape drives. Soon to come were a stream of high-speed paper tape readers, magnetic tape transports, line printers, servomotor tachometers, card readers, data terminals, card punches, disk drives, magnetic tape certifiers, optical page readers, and electronic

FIGURE 4.9. Control Data Peripheral Equipment Division at 7801 Computer Boulevard, Bloomington–Edina, Minnesota, 1970. The need for large production facilities sent Control Data into the Twin Cities suburbs in the 1960s.

scales—to list the announced products just from 1960 to 1966. The peripherals division itself undertook seventy-three identified development projects during this period.

Even though the company's marketing people wanted the CDC brand to appear on each and every product, Tom Kamp convinced Bill Norris that the large volumes possible with OEM manufacturing were a means to achieve impressive economies of scale.[27] Because Control Data needed to manufacture its own memory units, for instance, it might as well manufacture them in quantity. On disk drives shipped in the early 1960s, a breakthrough product for mass storage, Control Data undercut IBM's equivalent selling price by more than half, while the large volume drove manufacturing costs down and profits up. (Each disk drive unit cost Control Data roughly $3,000 to make, while selling it OEM for $9,000 or more, compared with IBM's sales price of $26,000.) With these attractive prices, Honeywell signed a $20-million contract for the CDC–OEM disk drives, while General Electric signed up for $35 million. In 1966, at last appreciating that such a valuable business deserved publicity, Control Data shifted its product announcements to accent their distinctive uses by a San

Francisco bank, the evangelist Billy Graham, and in the Autodin military communications network.[28]

Two other markets, in addition to the lucrative OEM business, helped Control Data grow peripheral products into a major corporate division. The dominance of IBM in the computer market led a number of would-be competitors to offer "plug-compatible" models that would match or exceed an IBM model's performance while undercutting its price. Honeywell offered a line of computers with the portentous name of "Liberator" aimed squarely at the IBM 1401 (see chapter 5), while RCA offered models to entice users from the 1401 model and, later, also IBM's System/360. Further entries by Raytheon, Amdahl, Fujitsu, and Hitachi created a huge market for "plug-compatible" peripherals. Because each of the manufacturers had designed their computers to be plug-compatible with the IBM world, they were also plug-compatible with Control Data's peripherals.[29]

The rise of minicomputers created a third market that transformed Control Data's peripheral products into a true mass commodity. The CDC 160 and 160A models selling at sixty thousand dollars were, on reflection, early instances of low-priced minicomputers, but it is not clear that the company recognized their full potential. In the mid-1960s, with much of the computing world's attention fixed on the upper-end "mainframe" computers, such companies as Digital Equipment and Data General created a new market niche for "minicomputers," an echo of the decade's fascination with miniskirts, that were built around low-priced solid-state electronics and bottom-end technical specifications. Prices tumbled. DEC's flagship PDP-8, first shipped in 1965, was initially priced at an astoundingly low eighteen thousand dollars. Although minicomputers were initially aimed at process-control applications in automated factories, they soon found their way into offices and laboratories, as well as into many diverse applications such as theatrical lighting, agriculture, and medicine. (Chapters 5 and 6 examine Honeywell's and IBM Rochester's respective forays into this market segment.)

Minicomputers created tremendous opportunities across the computer industry. They also created tremendous risks. "Within two years of the formation of Prime Computer [in 1972], they were trying to negotiate an order for 1,000 drives with CDC when they didn't even have any credit rating," Kamp remembered. An immense market for "miniperipherals" for these machines emerged between 1968 and 1972 when no fewer than one hundred new minicomputer models were put on the market. Control Data designed disk drives for the smaller machines, produced at a new factory in Omaha, Nebraska, as well as hit on a massively popular alternative to IBM's proprietary Winchester disk drive with its more simply designed Storage Module Drive (SMD). SMD was perfectly tuned to the minicomputer market, and this product line, across its life span, generated corporate revenues of $600 million.[30] This figure compares favorably with the revenues from the high-profile supercomputers.

FIGURE 4.10. Control Data 3000 series assembly, 1962. Early assembly methods ("one at a time out of a crib," at far right) occurred at McGill Building in downtown Minneapolis.

Peripheral products even helped transform the manufacturing of computer systems. Early on, all of Control Data's computers were assembled downtown "one at a time out of a crib, from a kit," and for years its top-of-the-line computers remained essentially one-of-a-kind craft constructions.[31] But that mode of manufacturing broke down when the backlog of orders for CDC's lucrative middle-line 3000 series grew to require six or seven deliveries each month. "So pretty soon Marketing has orders and they can't get deliveries. The factory can't produce. They are bogged down," recalled Tom Kamp. The problem was that each machine required a large number of components, making serial assembly "from a kit" extremely complicated. For example, the mid-level CDC 3600—priced at $1.2 million for a basic system in 1964—required 75,000 diodes and 52,000 transistors assembled onto two-inch circuit boards for its memory, compute, and communications modules as well as data channels. In addition, each 16 KB memory module required 852,000 tiny magnetic donuts, woven into wire frames, and fit into a cabinet that was roughly 3 feet wide, 6 feet high, and 18 inches thick. Add to this power supplies, controls, and assorted sockets for peripherals such as card readers, printers, paper tape units, or satellite computers. To deal with the frustrating factory backlog, Kamp took up a new position as vice president of manufacturing for eighteen months, bringing volume-production methods to the computer system division headed by ERA–Univac veteran Frank Mullaney.[32]

FIGURE 4.11. Control Data manufactures disk units, 1967. Computer manufacturing created thousands of jobs in Minnesota. By the mid-1970s Minnesota employed 17 percent of the nation's total computer manufacturing workforce.

Data Centers to Computer Services

Today, when such high-tech giants as IBM and Xerox have embraced the profitable and trendy world of "computer services," shedding their past glory in big iron hardware, it may be difficult to recall that computer companies once just sold computers. Some computer companies such as IBM were always closer to customers and hence to their needs for computing power to solve problems—not merely computer hardware. As early as 1957, IBM set up a wholly owned subsidiary called the Service Bureau Corporation to provide payroll, data processing, and time-sharing services to a wide range of client companies. Such services often depended on data centers, separately organized from the parent company's manufacturing and marketing efforts. The recent enthusiasm for "cloud computing" is a latter-day incarnation of data centers, with central installations of computers wired together into far-flung networks to deliver computing power as well as access to centrally stored data.

Control Data entered the data services business following the precedent established by ERA when it had set up its service bureau in Arlington, Virginia, with the hopes of selling time on an early ERA 1101. Offering only a computing machine and a few programs, however, ERA's early service business did not take off, and the computer instead found its way to Georgia Tech as a corporate donation. In 1960, Control Data opened a data center in Minneapolis, claimed to be the largest in the world, selling spare computer time to General Mills, Northern States Power, Honeywell Aerospace, and other companies. It created a pioneering computer services center in the San Francisco area shortly thereafter. In 1963, Control Data created a Data Centers Division under the supervision of H. F. Bloom, with additional centers quickly to come in Los Angeles, Washington, Cincinnati, Chicago, Houston, Boston, and other U.S. cities. By the early 1970s, there were two dozen data or services centers around the country and additional centers overseas in Frankfurt and other business centers. Whereas it was Seymour Cray's job to design and manufacture the computers in the early days, as Richard C. Gunderson, a programming supervisor hired from Univac

FIGURE 4.12. Control Data's Palo Alto Data Center (1965–67) was located at 3330 Hillview Avenue in Palo Alto, California, near the Palo Alto Programming and Development Center. Control Data tapped California programming talent to complement its Minnesota-based manufacturing.

who headed the flagship Minneapolis data center, recalled it, "my responsibility was to build the support organization, and that included software development; it included engineering services, and customer engineering training."[33]

For software, Control Data found that it needed to mobilize resources far beyond Minnesota. Its facility in Palo Alto, California, evolved into a major programming, software development, and services center. Already in 1960 Control Data had opened a sales office in nearby Sunnyvale, and CDC soon installed computers at the University of California–San Diego, General Dynamics–Pomona, Sylvania Electronic Systems–Mountain View, and the University of California–Berkeley, as well as the navy's Pacific Missile Range Facility in Hawaii. Richard "Dick" Zemlin, an ERA–Univac veteran then at Standard Oil of California, soon arrived to take charge of programming. Even then, California was a prime location for hiring computer programmers owing to the California aerospace industry.[34] Control Data even experienced difficulties hiring Minneapolis programmers for a post-sales contract with Lockheed's satellite test center, an early CDC 1604 customer. "When they noticed how easy it was to hire people to work on the Lockheed contract in the Bay area, they thought, 'Well, maybe we can build up the capability there,'" Zemlin recalled. For a time, CDC even ran a second software development facility in Los Angeles that directly tapped aerospace industry talent. The two rival software groups offered rival operating systems, with Los Angeles working on an early time-sharing system, and Palo Alto developing SCOPE—delivered in 1968 and officially known as Supervisory Control of Program Execution but sometimes comically labeled as Sunnyvale's Collection of Programming Errors. Control Data also built computer assembly plants at Sunnyvale in 1968 and in Westlake Industrial Park, outside Los Angeles, a year later.[35]

Initially, computer services evolved organically from computer sales. Every new sale of a large computer triggered a small raft of hiring with a sales representative being assigned to that site for follow-up business. For each of these million-dollar machines, there was extensive corporate training and post-sales support. "Often it was a condition of the contract that one of our people would go there for a period of time" to train the customer's staff, Zemlin recalled, with the roles of support, programming, marketing, sales, and training thoroughly mixed together. Customers might even take up programming activities. CDC's leading customers, including Lockheed and Oak Ridge, formed an active user group, broadly similar to IBM's acclaimed Share user group, that hatched "a whole new operating system and a set of compilers, an assembler and all to go with it" called the Co-op Monitor. It was, Zemlin recalled, "a unified architecture for all the software . . . a set of interfaces that you could attach further things to." The Co-op Monitor "was the thing that made Control Data believable as an integrated hardware and software company." This novel effort in distributed and cooperative software development (anticipating today's open-source software) was also overseen by Zemlin's group.[36]

FIGURE 4.13. Control Data Palo Alto software development, 1962. From left: Gary Bronstein (?), Donn Parker, Don Lytle, Richard Zemlin, Fred Laccabue, Robert Price, Bill Rosenstein, Chalkley Murray, Robert Hambleton (?). Identification by Fred Laccabue in 2009. Price rose through the ranks to become chairman and CEO in the 1980s.

The patchwork of support activities was consolidated under Robert M. P_· Price was a programmer with valuable computing experience at the national] Lawrence Livermore and at Convair's ERA 1103 (see chapter 2), whom Z brought over from Standard Oil of California (San Francisco). He was name_/ ager of application services and given the job of coordinating the contract-s software support, customer training, and documentation. For this crucial activity, a step toward full-blown computer services, "we would draw on pe_/ all the software groups in the company," Price remembers. Mike Schum of Price's programmers, recalled that "the environment to work in was o having a good time" with "a very, very highly motivated group of peo_· typical workweeks of seven days and eighty hours. "It was a tremenc environment," he noted. "You didn't have a set of software tools . . customer had a unique set of software and our thrust at that time was set of those requirements." As Ray Allard, a Minneapolis-based pr₍

120 INNOVATION MACHINE

traveled the country for support services, put it, "we felt that we were doing big things—we were going to grow and be very successful."[37]

Control Data's services strategy, judging at least from IBM's shift to services in the mid-1990s, was about two decades early. Norris was an early, strong, and determined backer of data centers, even to the dismay of the engineering executives such as Mullaney and Cray, who wanted to focus the company's resources on designing, building, and selling top-market computers. To drive his point home, Cray, for some years, annually proposed outright selling Cedar Engineering. James Harris, later vice president of the data centers division, remembered that data centers were "one of Norris' sacred cows." "He's a very persistent man and when he decides that something is going to get a damn good trial, it gets a trial. . . . So whatever we [needed]

FIGURE 4.14. Control Data's Minneapolis Data Center in 1974. Data centers and computer services formed one of Control Data's three strategic pillars, along with supercomputers and mass manufacturing of computer peripherals.

to make it successful we got. And in fact we got a few things we didn't even ask for." Corporate accounting favored the data centers, according to Harris. Outside of the executive ranks around Norris, however, for some years "we were the corporate orphans." Charles Crichton, head of the Minneapolis data center, noted that early data centers were mostly "the stepping stone to systems sales and leases . . . It was 1965 or 66 before I heard public announcements from Norris and others that service was going to be a big thing."[38]

Three strategic acquisitions between 1963 and 1968 signaled that Norris intended to use computer services to wean Control Data from its dependence on selling computer hardware. These sizable acquisitions, in addition to smaller ones that rounded out CDC's capacities in control systems, dramatically expanded CDC's capacities in computer and information services. First, in 1963, CDC bought the computer division of Bendix, gaining its talented staff and expertise as well as commercial divisions or corporate relationships in Canada, Mexico, and Japan where Bendix was aligned with the up-and-coming printer manufacturer C. Itoh. Bendix also provided CDC an entry into computer maintenance services, which gained importance with the sales of more than two hundred of its compact 160 and 160A minicomputers that needed the post-sales support that Bendix had already mastered. Second, in 1967, CDC purchased C-E-I-R, Inc., a unique research entity originally formed as a RAND-like arm of the U.S. Air Force (officially the Council for Economic and Industrial Research) in 1952 to identify economically significant targets in the Soviet Union, such as coal mines, aluminum plants, and steel mills, and then reorganized as a private corporation two years later under Herbert W. Robinson. In addition to selling computer time and its own proprietary programming, CEIR also developed expertise in using computers in such application fields as electric utilities, air-traffic control, petroleum refining, oil-tanker scheduling, and aircraft maintenance. The CEIR acquisition brought into the CDC fold its expertise in time-sharing computing as well as several services companies such as Automation Institute, which ran a set of programming schools in San Diego, San Francisco, San Jose, and Chicago, and a television-rating agency that subsequently became Arbitron.[39] And, third, in 1968 CDC bought Commercial Credit Corporation, which amounted to a large Maryland bank, with the aim of using it to finance the leasing of computers as IBM had long done. In 1970, Norris set up Services as a separate top-level division under the leadership of Robert Price, who later succeeded Norris as Control Data's president and CEO.[40]

By 1970, Control Data's Cybernet network, created two years earlier, had expanded to deliver computing services across the country. Anchoring the network were seven top-model CDC 6600s located in Palo Alto, Los Angeles, Houston, Minneapolis, Washington, New York, and Boston, connected through the company's own wideband communication lines. "We set it up in Chicago so that the customer could come in and, as far as he was concerned, the [CDC mainframe computer] might be in the

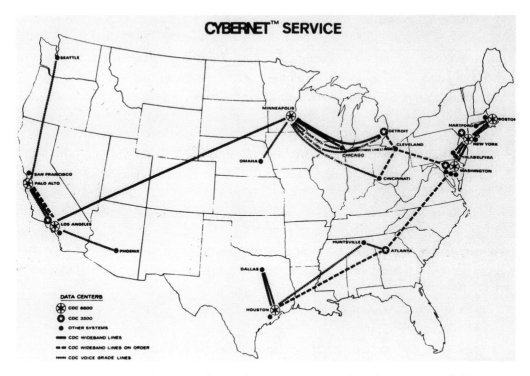

FIGURE 4.15. Control Data's Cybernet had as many network nodes in 1970 as did the ARPANET, the direct precursor to the Internet. Customers could submit their work locally, have it run on a Control Data supercomputer in another city, and pick up the results locally.

back room there, he didn't have to worry about it." The job was loaded into a small local computer in Chicago, then transmitted by telephone to a second one in Minneapolis that created a magnetic tape to be run on the central CDC mainframe; for output, the process was simply reversed, with the results arriving back at the customer's location.[41] In addition, there were middle-line CDC 3300s in Detroit, Hartford, and Atlanta, as well as nineteen other CDC computer systems in Seattle, Phoenix, Chicago, Philadelphia, Cleveland, and other cities connected through proprietary wideband or voice-grade communications lines. By comparison, the ARPANET, forerunner to the Internet, was in 1970 just expanding from its first four nodes on the West Coast to reach the East Coast with a link to Bolt, Beranek and Newman in Cambridge, Massachusetts. A famous map of the ARPANET in September 1971, reproduced in many places, identifies eighteen network centers, which compares with CDC's twenty identifiable network centers.[42] Cybernet soon added centers in Portland, Pittsburgh, and Rochester, New York.

Services received a massive shot in the arm following Control Data's epic lawsuit against IBM, launched in December 1968 and concluded in 1973. To tell a brief version, the suit resulted from IBM's sharp practices in "announcing" a special superfast model of its IBM System/360, which, even though it remained on the drawing boards for years, had the effect of hampering sales of Control Data's 6600 supercomputer. IBM at the time was also contesting its long-running antitrust lawsuit brought by the U.S. Justice Department (1969–82). Bill Norris was determined to settle the score. He charged IBM with a full range of unfair monopolistic practices, and demanded triple damages for "substantial and irreparable" losses. After years of legal wrangling, IBM agreed to a pretrial settlement, selling its Service Bureau Corporation to Control Data at a bargain price. "Acquiring the Service Bureau, which last year earned $1.5 million on revenues of $63 million, will swell Control Data's profits, as well as make the company a power in the services business," wrote *Time* magazine when the settlement was announced in January 1973. "Control Data now has 27 service centers of its own and will pick up 44 more from the Service Bureau."[43] And while the legal settlement required destruction of its database and all of "IBM's damaging documents," Control Data used this experience to launch a new services business in litigation support.[44]

Supercomputing Saga

In 1961, at an otherwise ordinary Monday staff meeting in downtown Minneapolis, Seymour Cray dropped a bombshell by announcing that he'd bought forty acres of land in his hometown of Chippewa Falls, Wisconsin, for a new house. When someone wondered about his unusually long drive to work, around one hundred miles, Seymour didn't miss a beat. "I've also taken an option on 40 acres of land next door to buy a plant." The executives were dumbstruck. There had been no discussion of the matter, and even Bill Norris was thrown off balance. After a long and awkward silence, Cray stated, "Well, somebody will build me one there."[45] Remarkably, of course, Control Data did just that. No four-year-old company in its right mind, especially one that was still juggling financial receivables, pitching for government projects, and trying to launch itself in the industry, would have allowed its chief engineer, no matter how much of a genius, to build a new laboratory-factory that was purposely removed from contact from corporate headquarters. But then Cray was not an ordinary genius.

Frank Mullaney, his boss from ERA–Univac days, told the following story about Cray. "He came into . . . the Atlas II job, . . . when we were quite a ways along and he hadn't been there more than a week or two and we had a prototype going, electrostatic storage. I saw a bunch of chassis out on the table and by gosh they were changing a component in all of these things. I said to Tom Rowan, who was in charge of it, I said, 'What are you doing?' And he said, 'Well, Seymour thought this ought to be changed.' Seymour wasn't really working on that, but Seymour thought this ought to be changed. They had already recognized him as somebody that they should listen to and he'd only been there a couple of weeks. That's sort of typical. He got into

FIGURE 4.16. Strutwear Building at 1015 South Fourth Street in Minneapolis (here in 1961) was the site of Seymour Cray's early design work on the CDC 6600 supercomputer, prior to his move to Chippewa Falls, Wisconsin.

everything."[46] Cray, officially, was responsible for designing the program control system for Atlas II, which became the successful ERA 1103.

Cray designed computers, at least in the early days, with his hands still gripping a soldering iron. "You have to remember that in those days, you didn't buy circuits from Motorola and Fairchild," as one of his engineers put it. "They were made up of discrete parts that you could buy during the noon hour in a brown bag, and then you would wire them together and make the circuit on the bench by yourself and it was a highly individualized thing." Cray's team possessed insight into top-level design combined with "the hands-on experience of how circuits work and how signals look when they go through cables and so forth."[47] The soldering iron and oscilloscope was just one part of Cray's tool kit. He had absorbed circuits theory and binary mathematics at the University of Minnesota, successfully designed the control system for the ERA 1103, created the prototype for Univac's transistorized NTDS, and then at Control Data had sold the U.S. Navy on his first transistorized commercial computer. Even though the mythology surrounding Cray typically singles out the ERA 1103, the Univac NTDS was the real training ground for the unusually talented engineers and managers that came to Control Data, including James Thornton, Les Davis, and Cray himself. Initially in Minneapolis, and then at Chippewa Falls, his efforts resulted in the famous Control Data 6600.

Cray literally was an architect. His eye synthesized a million details into a coherent whole, but he depended on his closest associates—James Thornton, Les Davis, Neil Lincoln—to translate his visions into practical problems that could be assigned to his design team. Thornton, as Cray himself readily acknowledged, was the codesigner of the breakthrough CDC 6600. Les Davis "didn't have a college degree and so he felt comfortable in that one step behind position . . . in the supportive role his contribution was as great or greater than mine," Cray recalled. "He dealt with all the personality issues, the human want issues, the needs of the people that were getting the work done. You can appreciate how important that was to me. He made things happen not only through his own work but through the works of others. Where I was allowed to go off merrily having my goal at the technical issues."[48]

Cray's oft-quoted statement about his ideal of designing computers starting from a "blank sheet of paper" has misled many commentators into an uncritical celebration of his genius, as if genius has no need for constraints and inspirations. As he told the Smithsonian's David Allison, "you have to understand that the blank sheet of paper is not a blank mind. I wanted to take advantage of all the things that I remembered and all the inputs I had gotten from people over a period of a few years to help me to decide what to make." It is important to remember that Cray had a strong background from the University of Minnesota as well as at ERA and Univac (chapter 3).

Cray is famous to computer engineers for a string of technical innovations such as RISC processing, vector processing, and high-speed packaging. Several of these

FIGURE 4.17. Seymour Cray in 1968 with CDC 7600 computer logic module. For many, Cray was the iconic computer designer; his supercomputers dominated the field for two decades.

innovations subsequently passed into the mainstream of computer design and are used by millions of computer users today. For instance, RISC processing—reduced instruction set computing, although some wags have dubbed it Really Invented by Seymour Cray—is used in smart phones, tablet computers, and numerous video-game platforms, as well as many high-end supercomputers. He also invented novel schemes—using such exotic substances as artificial blood for cooling—for getting rid of the immense amount of heat generated by his computers. Dimitry Grabbe, a mechanical engineer and pioneer in circuit board design, gave him this unusual but insightful compliment: "Seymour Cray, a brilliant person, I found no problem communicating with him. He was brilliant. He would say something, he would state it clearly. He understood what I was saying from half a word. . . . He—more then anybody else that I met—had understanding, appreciation, and insight into thermal behavior of parts."[49] One of Cray's recurrent jests was that he was an "overpaid plumber" because so much of his high-performance computer design dealt with pipes for cooling them.

The CDC 6600 brought world renown to Cray and a small mountain of cash to Control Data. Even IBM took notice, especially since its multimillion dollar Stretch computer was knocked off its pedestal. "Last week, Control Data . . . announced their 6600 system. I understand that in the laboratory developing the system there are only 34 people 'including the janitor.' Of these, 14 are engineers and 4 are programmers. . . . To the outsider, the laboratory appeared to be cost conscious, hard working, and highly motivated. Contrasting this modest effort with our vast development activities, I fail to understand why we have lost our industry leadership position by letting someone else offer the world's most powerful computer." So wrote IBM's legendary CEO Thomas J. Watson Jr. in a memo to eight of his top executives on August 28, 1963.[50]

It was said that Seymour Cray knew each of his customers by first name, quite possible given the small market for the "world's fastest" supercomputer. For instance, the National Security Agency was the lead customer for the ERA computers in the early 1950s, while the national labs at Los Alamos and Lawrence Livermore competed vigorously for the IBM Stretch and Univac LARC in the late 1950s. One way or another, Cray's computers dominated supercomputing from the mid-1960s through the mid-1980s. Owing to the fog of classification, we may never precisely know the contribution of the intelligence agencies to computing, but a comprehensive list of the first thirty-six systems at Lawrence Livermore gives a reasonable sample.[51] While IBM mainframe computers completely dominated Livermore's top computing muscle early on, apart from a Univac I delivered in 1953 and the LARC in 1961, from 1963 to 1982 it was Cray's exclusive show. Livermore bought two CDC 3600s, four CDC 6600s, five CDC 7600s, two CDC STAR 100s, and no fewer than seven of the iconic Cray-1s with their distinctive "circular love-seat" design. It would take a full month of continuous round-the-clock operation for a Univac I, assuming an exceptionally lucky run beyond the "average error-free running period" of around eight hours, to equal what the Cray-1 could compute in thirteen seconds.[52]

We now understand that the influence of the nation's nuclear weapons laboratories extended deep into the hardware and software of supercomputers. As historian Donald MacKenzie has noted, the deep-pocketed weapons laboratories encouraged computing development not along a single technical trajectory but along two distinct paths: some calculations were brute-force number-crunching "mesh" operations that required massive computing muscle, while other simulations using probabilistic Monte Carlo methods developed at the Los Alamos National Laboratory by John von Neumann and Stanisław Ulam required the computer to rapidly switch between different subprograms. Each path implied specific software and hardware. Cray himself identified a specific "multiply-and-increment" instruction that was designed to speed the immense "mesh" calculations done by the nuclear weapons laboratories. This computational technique was closely related to "vector" or "array" processing, where a computer was designed to work on arrays of data elements rather than on single data points. Vector processing was most famously implemented with the iconic Cray-1, which could work on eight vector registers each with sixty-four data elements. Such vector processing, ironically, is a departure from the pioneering von Neumann architecture that specifies orderly cycles of fetch–execute–store. Today, the Cell microprocessor—found in SONY's PlayStation 3, Toshiba's high-definition televisions, and IBM's Roadrunner supercomputer—uses a version of vector processing to speed graphics calculations.

Looking at the Minnesota-based ERA and CDC computers, it's clear that national intelligence and top-secret cryptography sent supercomputing down a third identifiable path. As one computer science source puts it, "the 0th [i.e., before even the 'first'] version of any powerful new computer, of course, always goes to the National Security Agency (NSA) for cryptographic applications." Well, of course. "The first of any new faster CDC machine was delivered to a 'good customer'—picked up at the factory by an anonymous truck, and never heard from again," stated another insider.[53] For years, the impact of cryptography and intelligence on computing was obscured. A good example is "A Nonarithmetical System Extension," chapter 17 of a semiofficial book on IBM's 7030 Stretch that includes oddly contrived examples from the "soft sciences," whereas the chapter—according to an unclassified NSA history—in fact describes NSA's HARVEST, a massive, one-of-a-kind hardware extension to IBM's Stretch designed for automated text matching of decrypted messages.[54] In the 1950s, as noted in chapter 2, NSA specified unusual "vector add" and "random jump" instructions to assist ERA's pioneering Atlas with cryptography as well as several "modular arithmetic" instructions for its successor.[55] The NSA itself, in a tantalizingly redacted history, states that "by the late 1960s, it is likely that NSA, with about _____ of equipment, had the largest collection of advanced computers in the United States, and probably in the world."[56]

One needs to look carefully, but this influence persisted across the CDC and Cray-era machines as well. The 1604's optional instruction number 77—for those models delivered to the intelligence agencies—turned it into a cipher machine, using a binary string as a key that activated a set of instructions and operations that, supposedly, "doubled the god damn repertoire and started all the fancy things over again . . . it was weird," according to Control Data engineers who worked with the machine.[57]

"Every Cray machine seemed to have a 'Population Count' instruction, which counted the 1 bits in a word—we figured this was for one customer—NSA," observes military computer expert Ed Thelen. "National Security Agency is charged with code cracking, and those who claim to know say the number of bits in a field is interesting."[58] Indeed, the significance of such a "population count sideways add" instruction directly owes to the work of information theorist and cryptographer Richard Hamming (1915–98) on the "weight" and "distance" of binary numbers.[59] The binary representations of numbers result in differing computational demands. While in decimal representations the numbers 31 and 32 appear to be only one digit removed, in their equivalent binary representations, 011111 and 100000, the "distance"—computed as the count of differing binary 1s and 0s—is much larger. And, to put things simply, with its five 1s the binary "weight" of 31 is 5, while the binary "weight" of 32 is just 1. Even today, the fantastically immense numbers used to encipher data for the RSA algorithm for public-key encryption, commonly used for electronic commerce and secure Internet traffic, are chosen with their binary "weights" in mind.

In the Community

We can now appreciate that Cray's supercomputers and Norris's leadership were essential parts of building a billion-dollar-a-year company. The mass manufacturing of peripheral products, innovative computer services, and lucrative supercomputers—together—were the engine that powered Control Data's innovation machine. It made a distinct impression on the state of Minnesota. One of its most visible community initiatives responded to the nation's urban distress in the 1960s. In the summer of 1967 racial riots had infamously torn apart Newark and Detroit—as well as the economically depressed north side of Minneapolis, where years of "white flight" had left poor black families socially and economically isolated. Control Data stepped into the breach. "There will soon be a new name on Plymouth Avenue, the street of boarded-up buildings . . . an exciting and significant development for our community," editorialized the *Minneapolis Tribune* in November 1967, announcing Control Data's decision to employ 270 people in a brand-new 85,000-square-foot urban factory. The *Tribune* welcomed Control Data's making a key computer component—the refrigerator-sized controllers that sent data to printers, storage units, and the like—signaling that the venture was not "some kind of corporate do-goodism or special federal contract." The *Minneapolis Star* joined the chorus of praise in "Industry Moves to the Ghetto," suggesting the plan was "based on hard-headed economics" that connected a computer factory and training institute with a pool of potential employees.[60]

The high-profile inner-city investment, soon to make Bill Norris an apostle of "corporate social responsibility," came at a favorable moment. Control Data's share price quadrupled in 1967 alone (it had previously slid from its early 1960s peak), and the company topped *Newsweek*'s list of high performers for the year. Evidently, the

FIGURE 4.18. Control Data's Northside factory, 1973. Control Data built facilities in depressed areas of Minneapolis and St. Paul that brought jobs to inner-city residents and favorable publicity to the company.

company was a stock-market darling that could do no wrong. Around this time, Norris directly addressed such "major social problems" as employment, health care, and education. He later told an interviewer that economist Milton Friedman's view that companies ought only to concern themselves with the "bottom line" of making profits for stockholders and furnishing goods for society, characteristic of times past, was simply no longer correct. Nor, as he told the interviewer, was his company pushed by antiwar protesters or corporate critics. "I think the greater motivation was recognition that these problems are being unsolved, and were threatening the very fabric of society."[61]

A full assessment of Norris's evolving business philosophy during the tumultuous 1960s remains to be done. To a striking extent, he was initially guided by the "hardheaded economics" noted by the *Tribune*. A month before announcing the Northside plant, he wrote to one of the company's founding stockholders, observing that "we are under tremendous pressure to increase minority employment."

> The government is going to spend a large amount of money to try to educate minority groups. It is Control Data's plan to get at the right end of the funnel when this money starts flowing, as it will eventually. . . . I am convinced that there is a big funnel in this government training thing, and if we put a plant in north Minneapolis along with a Control Data Institute, and show success, we will have a hose right on that funnel. Already there are a few politicians who have heard about our project plans, and they are anxious to get keyed in so they can get some credit. . . .

And soon enough—as if on cue for getting credit—Senator Walter Mondale delivered $1 million in Department of Labor job-training funds to the Northside plant. Control Data had perfectly positioned itself to "get the [federal] money flowing." For several months, Norris used the Northside investment—in rather narrow instrumental terms—to demonstrate that Control Data was already dealing with "the ghetto problem" and to deflect government officials' continual "request for a contribution." He noted that "with our North Side project, we are able to tell everybody with a great deal of sincerity that we are in it up to our eyeballs and there is nothing left."[62]

A decisive shift in Norris's thinking came in the wake of the assassination of Martin Luther King in April 1968, according to historian David Hart. As Norris explained to one of his executives—in a letter about a NATO industrial advisory committee (of all topics!): "It is my deep conviction that the poverty and riot situation has gotten so bad that it is by all odds the No. 1 problem for Control Data. We are in the fortunate position of having a little bit of a jump on the situation, but Control Data has to do much, much more." Not everyone agreed with the "realities of this situation," but in the weeks that followed, he engaged his top executives with the urban and political crisis. All sixteen of them were drafted to an evening lecture by noted Columbia University labor economist Eli Ginzberg, and Norris (in a private letter to an employee) resolutely defended Whitney Young, president of the National Urban League. As he told the employee, "we must listen to his views about the plight of the Negro . . . the stark fact of life may be that we are on the verge of anarchy."[63] Letters to such varied figures as the governor of Minnesota, a Chicago bank chairman, and a New York investment house each suggested that Control Data was "rethinking our business strategies . . . to help . . . with the poverty and crime problems."[64]

And yet, despite its urban investments and wide-ranging community activities, Control Data remained a thoroughly *suburban* company. Of its fifty facilities in Minnesota, totaling nearly 4 million square feet of office, factory, and warehouse space, and employing almost twelve thousand persons in 1975, almost all were located in the Twin Cities suburbs of Bloomington, Edina, St. Louis Park, Arden Hills, and

FIGURE 4.19. Control Data's fourteen-story headquarters was built in 1970 on a site just east of today's Mall of America. This modernistic building near the Bloomington Central light rail station is now headquarters of Minnesota-based HealthPartners.

Roseville (see Figure 4.20). Smaller facilities were sited in St. Paul around the Selby Bindery and at the Minneapolis Northside plant; in addition, the company opened training schools in Minneapolis and St. Paul (creating a total of 999 urban jobs or roughly 8 percent of the company's statewide employment). Smaller factories were already in the outlying towns of Cambridge, Montevideo, and Redwood Falls (roughly 3 percent of the company's statewide employment). At the time, Control Data, with nearly nine workers in ten employed in the Twin Cities suburbs, was a major force in suburban development, contrasting with the more urban-based Univac (see chapter 3). Its impressive glass-fronted corporate headquarters in Bloomington, built in the early 1970s, alone employed 2,600.

Across the 1970s and 1980s, Control Data evolved into a highly unusual computer company. It fought and won an antitrust lawsuit against industry giant IBM, strengthening its portfolio in computer services. It bought and then struggled to manage a Maryland bank. And it remained uncommonly active in a dizzying array of social, political, educational, and collaborative ventures. The company's leadership in founding the Microelectronics and Computer Technology Corporation, a pioneering computer R&D consortium, anticipated the cooperative SEMATECH consortium; both were located in Austin, Texas. The company's Rural Venture, City Ventures, and Fair Break initiatives reflected Norris's vision of creating business opportunities to deal with "society's unmet needs" such as job readiness, education, health care, and prison reform. Control Data's education initiatives include its CDI training schools, the innovative PLATO educational platform, and support for the Minnesota Educational Computing Consortium (see chapter 7). Research at the University of Minnesota handsomely benefited from the Microelectronics and Information Sciences Center, co-funded with Honeywell, 3M, Sperry Rand, and the state of Minnesota. Norris's own Minnesota Family Farm Institute, Minnesota Business Partnership, and Minnesota Wellspring, among others, explicitly targeted the state of Minnesota. Through the 1980s, the company was lionized for its financial prowess and social vision. Control Data's full-time employment in Minnesota peaked during 1981–84 at more than twenty-five thousand.[65]

In the 1990s, however, Control Data's slippage from its glory years prompted much hand-wringing and finger-pointing. Even today, it is by no means clear what really went wrong. During the midst of the personal computer revolution, the company went through major upheaval, skirted insolvency, and was ultimately split up. Some have faulted Norris's expansive social activism and determined focus on PLATO-based education, and some of the outside ventures may have been a distraction.[66] There were significant challenges in running a large, diverse corporation. While peripheral products needed the attention of a manufacturing guru, the Commercial Credit subsidiary needed bank-style management, and the services businesses needed patient yet opportunistic management of a different sort—as IBM, HP, and Xerox have discovered in recent years with their own ventures in computer services. After Seymour Cray left to start his own company in 1972, Control Data's supercomputer effort lost its celebrity power, just when Japanese and other manufacturers turned supercomputing into a lower-margin commodity-like field. In 1993, Silicon Graphics announced that it had "packed the power of a multimillion-dollar Cray supercomputer into a line of desk-side machines starting at just $120,000."[67] No one could make supercomputer profits selling such machines.

Since Control Data's two successor companies have themselves disappeared, it may be difficult at first to see the company's legacy. But today you can drive out to Bloomington–Edina, take an exit off Highway 100 immediately north of Interstate

City	Total Area (1,000 square feet)	Employment
Arden Hills (4 plants)	878	2,004
Bloomington (corporate headquarters)	861	2,616
Bloomington (6 plants)	409	1,743
Normandale (original + additions)	381	1,570
Cambridge (1 plant)	13	120
Edina (7 warehouses/plants)	224	377
Eagan (1 plant)	41	168
Minneapolis (6 plants)	198	567
Montevideo (3 plants)	12	70
Plymouth (2 plants)	250	775
Redwood Falls (1 plant)	40	187
Roseville (5 warehouses/offices)	308	702
St. Louis Park (3 plants)	92	536
St. Paul (3 plants)	83	432
Total (50 facilities)	**3,790**	**11,867**

FIGURE 4.20. Control Data facilities, space, and employment in Minnesota, 1975. Source: Control Data Corporation, Facility Summary Report, January 1, 1975; CBI (80), series 21, box 2, folder 2. The 1975 report was the last one listing employment figures.

494, and find the aptly named Computer Boulevard. There is Seagate Technology, which built its successful hard-drive business around Control Data's high-end technology. A street sign for "Disk Drive" marks the Seagate plant's entrance. Control Data Systems became part of British Telecom, while the services business, initially spun off as Ceridian, eventually became part of Fidelity National Financial. Innovations from Control Data's programming legacy were taken up by today's software giants Adobe and Symantec. Anyone using a Citibank credit card also holds a bit of Control Data in their hands, because Commercial Credit was spun off to Sandy Weill, who merged it with several financial services companies to form one of the nation's largest banks. And if you are listening to the radio, Arbitron is watching you.

5
First Computer

Honeywell, Partnerships, and the
Politics of Patents

For years Honeywell was Minnesota's largest private-sector employer and a mainstay of its high-technology economy. It was one of several established American electronics companies, including General Electric and RCA, that launched themselves into the computing industry with the aim of capitalizing on their electronics savvy and manufacturing prowess. These three companies enjoyed dominant positions in radio, television, and industrial electronics and seemingly had the best chance of contesting IBM's rising power in computing. Honeywell was for many years entangled with General Electric, as chapter 7 makes clear. These two companies, along with researchers at MIT, jointly developed a pioneering computer time-sharing system in the 1960s; and by 1970 Honeywell took over General Electric's entire computing department. (Three decades later, General Electric tried to purchase the entirety of Honeywell, but that takeover was blocked by European antitrust regulators.) Also in the 1960s Honeywell launched its own independent corporate efforts in smaller, medium, and large-scale computing and struggled to make them successful.

Throughout these years, Honeywell's national profile in military technology expanded impressively. It became such a prominent symbol of the military–industrial complex—a concept coined by Malcolm Moos, speechwriter for President Eisenhower and later president of the University of Minnesota—that it caught the eye of antiwar protesters in the Twin Cities, who organized the Honeywell Project (1968–90), an unusually long-running protest movement.[1] In 1990 and 1991, Honeywell spun off its military activities as a separate corporation known as Alliant Techsystems Inc., and then sold off its computer division to the French conglomerate Groupe Bull. Eight years later, the core Honeywell concern itself was bought up by the corporate giant Allied Signal, which retained the brand name of "Honeywell International" but moved

136 FIRST COMPUTER

its headquarters from Minneapolis to New Jersey. It is a telling sign of the transformation of the state's economic base that the Honeywell complex just south of downtown Minneapolis is now the prominent home of Wells Fargo's home mortgage division. In the same buildings where a wide range of industrial, military, and consumer goods were made—including Honeywell's signature "round" home thermostat—mortgages are now manufactured.

A century from now Honeywell may be remembered as the most famous of Minnesota's computer companies. It is the only computing company mentioned twice in

FIGURE 5.1. Manufacturing Honeywell's signature "round" thermostat, circa 1955. Honeywell created a major industrial district in South Minneapolis. The "round" thermostat was among the 150 Minnesota icons recognized by the Minnesota Historical Society. Courtesy of the Minnesota Historical Society.

the Minnesota Historical Society's tabulation of the 150 events that shaped the state's history. Despite its ups and downs in computing, Honeywell had notable technical successes with the GE–MIT Multics time-sharing system, recently gaining appreciation as an inherently secure computer system; it injected some healthy competition into the computer industry with its provocative line of "Liberator" computer systems aimed at IBM; and its minicomputers formed the backbone for the ARPANET, the direct forerunner to the Internet. Its massive manufacturing complex in South Minneapolis was a fitting anchor to the urban industrial district that grew up around it. All the same, Honeywell's enduring fame will likely be for none of these computing accomplishments, but rather for its high-profile legal strategizing. It won a landmark federal lawsuit against the corporate parent of the local Univac division, the now-famous *Honeywell v. Sperry Rand* case. The lawsuit, a direct extension of the Philadelphia story, challenged the pioneering ENIAC patent that was inherited by Sperry Rand (see chapter 1). Sperry Rand claimed ownership over the concept of the computer and sought to enforce its patent rights through hefty royalty payments from each and every computer manufacturer. Millions of dollars hung in the balance. Honeywell launched a countersuit to defend itself, attacking the validity of the core ENIAC patent. The legal wrangling—in a Minneapolis courtroom—stretched out over years (and remember that the ENIAC was publicly announced in 1946). Finally, in 1973, the *Honeywell v. Sperry Rand* decision settled the legal question of who invented the first computer. Whether you sided with Honeywell or Sperry Rand, Minnesota was at the center of computing history.

Born Digital

The origin stories of Honeywell and IBM (see chapter 6) have a number of curious parallels. Both of these corporate giants had technical roots in key patents won by unconventional mechanical inventors of the late nineteenth century, both were formed through unusual corporate mergers, both thrived during the difficult days of the Great Depression; and each of them had early, extensive, and productive business with the U.S. federal government. While IBM mastered the tabulating machine industry and grew large organically, from internal revenues and investments rather than by purchasing other companies, Honeywell grew rapidly through multiple acquisitions, often driven by patent concerns, and came to dominate the wide-ranging field of industrial controls. It perfected the means for automatically controlling industrial assembly lines, chemical plants, and even complex electronics through linkages of sensors and sophisticated feedback mechanisms. Its signature product, the famous round Honeywell thermostat found in millions of homes, was in fact emblematical of the company's background and technical success.

When it took form in 1927, the Minneapolis–Honeywell Regulator Company was heir to two separate companies. The younger company, founded only two decades earlier in Wabash, Indiana, specialized in large-scale industrial heating. The Indiana purveyor of oversize industrial stoves was the brainchild of entrepreneur Mark Honeywell; and in fact it had begun to compete with an older and established Minneapolis company whose roots stretched back to the patents of Albert Butz. A prolific Swiss inventor, Butz (1849–1904) created what we can now appreciate as early feedback control devices. In 1884, three years after he arrived in St. Paul, he formed the short-lived Butz and Mendenhall Hand Grenade Fire Extinguisher Company to exploit a fanciful scheme using suspended glass globes filled with water or carbon tetrachloride to fight fires. If a fire broke out, something like feedback occurred in the sense that when the strings holding up the glass globes burned through, they fell to the floor and cracked open, supposedly putting out the fire. Evidence of this company's success is not readily at hand.

More successful than this dubious fire-extinguishing scheme was Butz's "damper flapper" patented the following year. At the time, many individual homes and buildings were heated by a coal-fired furnace. Residents went down to the basement several times each day, shoveling in fresh coal and clearing out spent ashes—as well as manually adjusting the draft of air. If the incoming draft of air was too strong, the furnace would overheat, wasting fuel and cooking the building's inhabitants; but, conversely, if the draft was not strong enough, the coal fire might die down and freeze the inhabitants. The damper flapper automated the airflow to the furnace. If the furnace's temperature got too high, the damper automatically closed down, cutting off the flow of air; and when the temperature fell too low, the damper opened up, increasing the draft of air and increasing the temperature of the coal furnace. It was a simple on–off device connected directly to the furnace. In a set of obscure dealings, Butz signed his patents over to a group of Minneapolis businessmen who, facing economic difficulties in 1892, in turn sold their patent rights for $1 to William R. Sweatt, an up-and-coming business promoter looking for the main chance.

Even after inventor Butz left for Chicago in 1888, opening his own shop there, something of his Swiss technical prowess, along with his valuable patents, stayed in Minneapolis. His mechanical savvy was one of the origin points not merely for Honeywell but also for the fledgling Minneapolis industrial district. As chapter 8 makes clear, there are important "forward" connections between the computing and medical device industries, and it is worth pointing out this early case of direct forward connections from the mechanical tradition to the computing industry. Sweatt had previously invested $5,300 in the fledgling concern, and he proceeded to build up the Minneapolis Heat Regulator Company into a manufacturing powerhouse. In 1912, Sweatt built the first installment of what became a mammoth manufacturing complex at Fourth Avenue and Twenty-seventh Street in South Minneapolis; today this is part of a Wells Fargo home mortgage complex, easily viewed along Interstate 35-W.

FIGURE 5.2. Interior of the Honeywell Relay Department, circa 1930. Precision manufacturing begun by Swiss technicians augmented the metro's high-tech industrial district. Photograph by Lee Brothers. Courtesy of Minnesota Historical Society.

Sweatt, along with his two sons as junior partners, recognized the vital importance of sales and marketing, and he prompted stove manufacturers to adopt designs that might easily be refitted with his patented temperature-control system. Advertisements in such national magazines as *Saturday Evening Post, Harper's, House Beautiful,* and *Good Housekeeping* in 1915 linked the Honeywell product to an image of comfort and convenience.[2] He also seized on the marketing advantages of a redesigned product that made it easy to attain, as a 1927 sales brochure put it—"the ideal temperature which all of us should strive to enjoy is 68 degrees"—while the house or building's owner remained all the while comfortably upstairs. The most technically sophisticated product of these years combined accurate temperature control with a seven-jeweled clock mechanism that automatically raised the house's temperature in the morning and lowered it at night. Production facilities capable of turning out forty thousand clock thermostats by 1927 required "tiny lathes and drill presses so fine and accurate in their construction that it was necessary to send abroad to get them."[3] These are further indications that Minneapolis was investing in the capability of sophisticated high-precision manufacturing.

Until the late 1930s, all of Minneapolis–Honeywell's furnace products were essentially digital switches that sent binary "on" or "off" commands to the motors, blowers,

FIGURE 5.3. Honeywell Building, Twenty-seventh Street and Fourth Avenue South, Minneapolis, circa 1955. In 1942 Honeywell expanded its north addition to the main plant (here at far left) to ten stories. This complex can be seen today next to the Wells Fargo Home Mortgage building. Source: C. W. Nessell, *Restless Spirit* (Minneapolis: Honeywell, 1963), 101.

and heaters that they controlled. In this sense the company was born digital. In 1939, Honeywell engineers began experimenting with regulator mechanisms that *varied* the amount of feedback control. If, for instance, the temperature was just one degree low, the correcting signal for more heat would be small, but if the temperature was several degrees lower than the target temperature, the correcting signal would be larger. Heated buildings, even in Minnesota's coldest winters, were not usually subject to severe temperature swings and did not need the novel continuous regulation system. Minneapolis–Honeywell needed to look elsewhere for a promising market. It is little remembered now but at the time there was a sizable airplane manufacturing cluster in Minnesota. Not only did Northwest Airlines link businessman John Parker to the manufacturing of computers (as told in chapter 2), but there were also Minneapolis-based Mohawk Aircraft Corporation, which hired as its chief designer John Akerman, who went on to found aeronautical engineering at the University of Minnesota, and Bellanca–Northern Aircraft of Alexandria, Minnesota. The latest generation of airplanes were subject to extreme temperatures—it might be ninety degrees on the ground while an airplane could face thirty degrees below zero at typical flying altitudes of the time—and the company's engineers sketched out a way of continuously mixing varied amounts of heated and nonheated air to automatically maintain a comfortable environment for the crew at different altitudes.

With the country's military mobilization beginning already in early 1941, even before Pearl Harbor, Honeywell engineers took their variable-feedback temperature scheme to Wright Field in Dayton, Ohio. The aircraft designers were impressed with the feedback scheme but not owing to any particular comfort concern for aircraft crews. At the heart of Honeywell's scheme was a variable amplifier useful for precisely controlling any servomechanism: it was the implementation of *variable* feedback control and not the specifics of temperature regulation that caught the attention of Wright Field's aircraft engineers. They had a different problem right at hand. The Army Air Force at the time was seeking some means of stabilizing its supersensitive Fairchild surveillance cameras while they were taking precisely positioned photographs of the terrain below. As it turned out, the Honeywell variable-feedback control system was very good at stabilizing and controlling the mechanical movements of the heavy surveillance cameras—and, before long, much else on airplanes besides.

Even as Honeywell's high-precision mass-manufacturing during the war years churned out a total of 4.7 million special military devices such as tank periscopes, gunner's quadrants, mortars, and other artillery, a dividend from the Swiss-inspired precision manufacturing the firm had absorbed from inventor Albert Butz, the firm's engineers extended its variable-feedback control technology to entirely new realms. The most important of these was the development and deployment of the C-1 automatic pilot. Just as Honeywell's servomechanism worked to keep a surveillance camera level in flight, the same basic control technology was grafted onto the famous top-secret Norden bombsight to keep an entire airplane level in flight. The Norden bombsight was an ingenious mechanism, essentially a robust mechanical analog computer, designed to predict the instant "when" a bombardier should release the plane's load of bombs to hit the intended target below. This was no simple task. With bombers flying at altitudes of twenty thousand feet or more to avoid antiaircraft fire from the ground, it might take half a minute or more for a bomb to fall to the ground. During that time, a bomber would travel a full 1.5 miles on its flight route while a bomb, too, would cover most of this distance as it fell forward in a long arc toward its target.

The Norden bombsight mechanically computed the proper time for releasing a bomb load, taking into account the plane's forward speed and altitude as well as compensating for crosswinds in an attempt to improve accuracies. One serious drawback was that the Norden system presumed that the plane was flying perfectly smooth and level, and at constant speed, throughout the bomb run. Few pilots achieved

FIGURE 5.4. Precision mass-manufacturing at Honeywell plant, Minneapolis, 1945. Photograph by *Minneapolis Star Journal.* Courtesy of Minnesota Historical Society.

that nerve-wracking goal. Honeywell first built an add-on stabilizing autopilot to the Norden and then, to expand production volumes, it built an integrated unit that combined automatic pilot with bombsight. Honeywell manufactured an astounding thirty-five thousand autopilots by the end of the war.[4]

While the ultimate accuracy of strategic bombing during the war remains in some doubt—it was a wild exaggeration when enthusiasts claimed to be able to hit a "barrel full of pickles"—the automatic pilot is widely acknowledged to have worked wonders. The Honeywell automatic pilot smoothed out not only the troublesome bombing runs but also the rest of the flight. It became common to have flight reports of an entire sixteen-hour flight done on automatic pilot, except for brief periods of takeoff and landing where pilots were required to take up the active control of the aircraft. Once in the air, pilots could tweak the unit's knobs to alter the plane's direction, speed, or altitude, all the while keeping the plane steady under automatic control. One pilot of a shot-up plane, struggling to regain control of the heavily damaged aircraft, found success at stabilizing it when he switched on the automatic pilot. One

FIGURE 5.5. Norden bombsight, 1953. Honeywell manufactured thirty-five thousand autopilots and thousands of the Norden bombsights en route to becoming a major military contractor.

hapless crew on a mission over North Africa ran dangerously low on fuel over a sandstorm and was ordered to parachute down; most of the crew members hit the ground in the snarling winds and, unable to get free of their parachutes, were dragged to death. The plane itself continued in a long slow descent until, as though perfectly timed, it ran out of fuel and gently coasted in for a landing on the desert.

One might suggest that Honeywell's distinctive system of variable-control feedback was a ready solution in search of new problems to solve.[5] During the war, Honeywell built on the successes of the pioneering C-1 autopilot. Soon similar Honeywell control units were installed on B-17, B-24, and B-29 bombers to control these planes' turbochargers, which pumped in massive quantities of pressurized air to keep the piston engines running effectively at high altitudes. As an airplane changed speeds and elevations, the Honeywell controls automatically adjusted the turbochargers to keep the engines at peak efficiency. Pilots who were already adjusting knobs on the C-1 autopilot also liked the "formation stick," which electronically linked their steering yoke to the autopilot, allowing the pilot to maintain formation or other close-drawn flight maneuvers with a light touch. During the war, Honeywell logged 1,800 hours evaluating its control equipment in a special-purpose B-17 test airplane. During the subsequent Cold War years, Honeywell's high-technology military business boomed. Honeywell contributed sophisticated control equipment to something like 90 percent of all orbiting space vehicles, including "nearly every major civilian and military space effort." Civilian aircraft were not far behind. "When the [Boeing] 707s back around 1960, began flying transatlantic flights, every company that had something to do with the 707, such as Honeywell with the electronics and all that, really took advantage of that as a publicity opportunity, and played it up big," noted one well-informed observer of Minnesota computing.[6]

FIGURE 5.6. Engineer Leonard Aske demonstrates the sensitivity of Honeywell's nine-ton spacecraft simulation platform in 1964. Honeywell built control systems for airplanes, missiles, spacecraft, and other high-tech applications. Photograph by Earl Seubert. Courtesy of Minnesota Historical Society.

Partnerships in Computing

Honeywell's entry into the computer market took form through joint ventures. Its first was with Raytheon in the mid-1950s. Raytheon had been cofounded in the 1920s by MIT electrical engineer Vannevar Bush (see chapter 1), and it became a major supplier of vacuum tubes for radios and microwave radar, which gave it a head start in microwave ovens. In 1948 Raytheon began manufacturing guided missiles, which prompted the company to build its RAYDAC computer (installed at the Point Mugu Naval Air Missile Test Center) as well as provided a ready and lucrative alternative when commercial computing proved too fickle. The Honeywell–Raytheon joint venture Datamatic 1000 was a classic mainframe computer of the time, weighing thirty-five tons, and with hybrid circuitry comprising 3,600 vacuum tubes as well as 60,000 semiconductor diodes and 500 transistors. Its chief distinguishing feature, according to one technical account, was its ability to read and write 60,000 decimal-digits per second to its three-inch-wide magnetic tape drive. Despite its stiff cost of $2 million or so, Datamatic 1000s were installed in 1957 by the Michigan Hospital Service/Blue Cross, and then by the Baltimore & Ohio Railroad, First National Bank of Boston, the U.S. Treasury Department, and at the headquarters of Honeywell itself, where the machine was used for payroll, accounting, inventory, and factory efficiency studies. A fifth machine, installed at Datamatic Division in Wellesley Hills, Massachusetts, became the core of a programming and training center that Honeywell ran for some years.

In 1960 Honeywell created its own Electronic Data Processing division.[7] Its efforts to wrest customers away from IBM led to the famous Honeywell 200 "Liberator," which offered some enticing gems to the industry. During 1965, a monthly series of advertisements for it in the leading trade journal *Datamation* promised "a fast getaway from a 1401," claiming that "your present [program] tapes will run without change . . . Our Liberator will turn your present programs into fast-moving new programs . . . up to five times faster than a 1401." At least one corporate executive found that Honeywell's machine performed as capably as a comparable IBM model—and at half the monthly rental. For its part, IBM fought back with its "IBM Captivator software that stymied Honeywell 200 Liberator conversions of IBM programs. It was a card box full of subroutines that could be used by Autocoder programmers for many functions. Honeywell's Liberator software could not handle these subroutines."[8]

Honeywell entered the hurly-burly minicomputer market in 1966 when it purchased the Computer Control Company, itself an earlier spin-off from Raytheon. Its guiding spirit was Louis Fein, the designer of the singular RAYDAC computer (1949–53) and later a prominent figure in computer science at Stanford.[9] The company was based south of Boston in Framingham, Massachusetts, and in several ways it paralleled the Digital Equipment Corporation, first designing components for digital circuitry and then manufacturing its own minicomputer. In 1963 it introduced a state-of-the-art

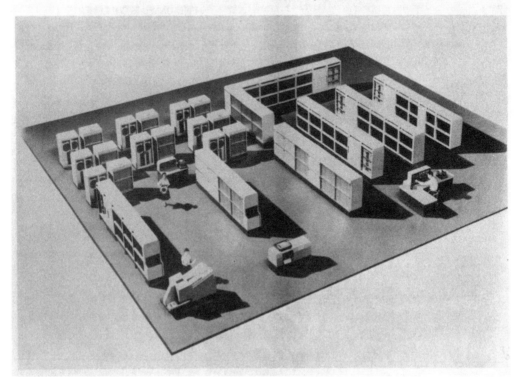

FIGURE 5.7. A joint venture between Honeywell and Raytheon resulted in the sprawling Datamatic 1000, circa 1960. The thirty-five-ton machine (with 3,600 vacuum tubes, 500 transistors, and 60,000 crystal diodes) was installed at the Baltimore & Ohio Railroad, the First National Bank of Boston, the Michigan Hospital Service, the U.S. Treasury Department, and Honeywell's Minneapolis headquarters. Source: Weik, *A Third Survey of Domestic Electronic Digital Computing Systems*, 214.

sixteen-bit minicomputer, the DDP-116, that found a ready market in process control applications and later in aerospace simulation. After being purchased by Honeywell, and turned into its Computer Control Division, it produced two justly famous computers along with a large number of embedded control computers. The Honeywell H316 was a reengineered successor of the DDP-116. "Most of the logic design is identical. . . . By using more highly integrated chips, and later generation core memory, it was possible to reduce the size from something the size and weight of a small car to something that fits under a desk and can (just) be lifted by one person."[10] Teaming with upscale retailer Neiman-Marcus, Honeywell offered a pedestal model of the 316 as a fantasy "kitchen computer." For a mere $10,600, including two weeks of

FIGURE 5.8. Honeywell's "kitchen computer" at the office. In 1969 Honeywell modified this pedestal-model $10,000 minicomputer for use in the home kitchen. After a two-week programming course, the housewife "by simply pushing a few buttons [can] obtain a complete menu organized around the entree." *Life* magazine published a photograph of this "unusual Christmas gift."

computer programming lessons, the modern housewife could use the computer to store recipes and plan menus: "If she can only cook as well as Honeywell can compute."[11] In turn, one might say that the successor Honeywell DDP-516 formed the first version of the Internet, because slightly modified versions of these machines were turned into the hardware nodes of the ARPANET (1969–89), the Internet's immediate precursor.

Minneapolis-based Honeywell, if it had played its cards differently, might well have dominated educational computing. In 1970 Honeywell bought up General Electric's entire computer division when GE decided to exit the business, as noted earlier. No

FIGURE 5.9. Honeywell's "coast-to-coast" computer network, 1969. Honeywell built a nationwide time-sharing computer network with Minneapolis at the center. The system "links Honeywell engineers from Boston to Seattle with . . . the automation company's Minneapolis headquarters."

one at the time was selling better time-sharing computers than GE's, and this immense resource fell into Honeywell's lap. (For years IBM struggled to achieve workable time-sharing on its System/360 computers that came to dominate the business world, but not until the mid-1970s was this goal truly achieved.) Honeywell also set up a national network—known as EDINET—for educational computing, with centers in Minneapolis, Atlanta, and San Francisco. To access the Honeywell computers, schools needed to purchase the usual teletype terminal, acoustic coupler, and have access via phone lines—and then on top of that write a monthly check to Honeywell for one thousand dollars in exchange for unlimited access. For many schools, this cost model proved to be unattractive; only wealthy schools with extensive computer activities already in place could really justify the hefty monthly fee, and conversely the flat-rate fee was difficult for any school just trying to get its own educational computing off the ground. In downstate Minnesota, Mankato State created an attractive work-around by sharing its EDINET account at Honeywell with twenty-five or thirty nearby schools in south-central Minnesota, thereby splitting the heavy monthly costs. It is likely that the prototype "Oregon Trail" computer game was written on this system; this game made the Minnesota Educational Computing Consortium literally millions of dollars (see chapter 7).

Honeywell, even though it had some early success in combating IBM with its provocative "Liberator" line of computers, was perhaps too focused on the central mainframe model of computing. Many school districts did the math and concluded that they could save money by outright purchasing a smaller minicomputer from Hewlett-Packard or DEC rather than paying those interminable monthly payments to Honeywell. With its own minicomputer division at hand, Computer Control Company purchased in 1966, Honeywell seemed ready to compete directly with minicomputers from DEC and HP. It directly tapped into the Minnesota educational network (chapter 7) by hiring several of its leading figures for the EDINET initiative. But the pieces, somehow, never gelled. As it turned out, Honeywell in a few years wrapped up EDINET, moved its computing business from Minneapolis to Massachusetts, and went whole hog after IBM in the business market. In the 1970s it attempted to centralize its computing efforts as Honeywell Information Systems, led for years by Clarence "Clancy" Spangle, but in 1991 Honeywell closed out this chapter by selling its computer business to the French-based Groupe Bull.

Honeywell v. Sperry Rand (1967–73)

The battle to claim the title of "first computer" has been intensely contested ever since the U.S. Army lent its publicity muscle to the ENIAC effort in 1946. The army's efforts had the effect of pushing aside the significant British efforts to build early stored-program computers at the universities of Manchester and Cambridge, as well as the wartime effort to automate code breaking with Britain's Colossus computing engine and, for that matter, the American work that resulted in ERA's Atlas computers. Also pushed to the side were the efforts of German inventor Konrad Zuse, who built mechanical computers with binary logic in the 1940s, as well as Bell Laboratories, which built several generations of computers with electromechanical telephone relays beginning in 1940. As Table 5.1 indicates, the variety of early computers was nothing short of astonishing.

It seems incomprehensible that the legal questions surrounding the invention of computing in the 1940s were not settled until 1973, nearly three decades after the fact. It is beyond the scope of this chapter to do anything like a definitive assessment of who invented the first computer, and quite likely this question will never be neatly resolved. When historians closely examine such seminal technologies as automobiles, electric lights, or radios, they typically find complicated, tangled, and openly competing claims about their invention. It is not only that inventors, and their patent attorneys, shaped the available evidence instrumentally to fit their preferences and priorities; it is commonly difficult to define "what" a specific technology consists of,

TABLE 5.1 Early digital computers, 1940s

Computer	Year	Data	Computing	Program Storage
Bell Relay	1940	Bcd	mechanical	(not programmable)
Zuse Z3	1941	B	mechanical	film stock
ABC	1942	B	electronic	(not programmable)
Colossus	1944	B	electronic	patch cables + switches
Harvard Mark 1	1944	D	mechanical	paper tape
ENIAC	1945	D	electronic	patch cables + switches
Bell Relay V	1946	Bq	mechanical	paper tape
Manchester Baby	6/48	B	electronic	stored: Williams tube
EDSAC	5/49	B	electronic	stored: 32 delay lines
BINAC	9/49	B	electronic	stored: 2 delay lines
NBS SEAC	5/50	B	electronic	stored: 64 delay lines
ERA Atlas	10/50	B	electronic	stored: magnet drum
EDVAC	circa 1951	B	electronic	stored: 64 delay lines
UNIVAC	1951	Bcd	electronic	stored: 100 delay lines

This table is composed from many diverse sources and contains numerous judgment calls, especially about dates. **B** indicates a machine with a binary-based CPU; D is a machine that calculated with decimal numbers; **Bcd** is a machine that used binary-coded decimal numbers; **Bq** is a variant of Bcd using so-called bi-quinary numbers.

especially one with many separate parts welded into a coherent system of technology. When asked to give his definition of a computer, one pioneer, reasonably enough, declined: "I can tell you about different kinds of computers, but I can't say what a computer is."[12] All grand claims about "the computer" strongly tend toward this familiar if somewhat frustrating state of affairs. We can clarify why the ENIAC patent was not issued until 1964 and make an appraisal of the landmark *Honeywell v. Sperry Rand* trial, which was by any measure truly a monumental undertaking. The courtroom trial was at 135 days the longest ever in the federal system, and it rested on more than thirty thousand trial exhibits, six years of pretrial depositions, while resulting in a court transcript of more than twenty thousand pages.

A persisting cloud hung over the ENIAC patent owing in part to Eckert and Mauchly's own commercial inexperience. They were both essentially academic researchers and would-be business entrepreneurs, and despite being computer visionaries they sorely lacked trustworthy counselors to help them navigate the shoals of successfully running a new company. Their first serious mistake, as noted in chapter 1, was failing to complete a patent application on ENIAC in a timely fashion. Their filing did not occur until June 1947—substantially, and fatefully, exceeding one year after ENIAC's public dedication in February 1946. The delay had severe consequences, as we will explore below. Another mistake can be seen more clearly in retrospect. Because the ENIAC application was contested by several rival parties—

various companies filed a total of eleven patent interferences claiming priority in invention—Eckert and Mauchly had extensive opportunity to amend their original application. With the able assistance of a team of patent attorneys, the document grew ever larger until by the time it was eventually issued in 1964, it reached an unwieldy 208 pages and contained an ill-advised 148 separate claims.[13] Having a single patent with only Eckert and Mauchly's names on it also meant that other members of the ENIAC team who made significant contributions were frozen out. While companies such as GE, RCA, and AT&T, well schooled in the radio patent wars, knew that an effective strategy was to build an interlocking set of "1,001 little patents and inventions" with separate but supporting claims, the ENIAC patent was one immense and fragile monolith.[14] Two rival companies exploited this weakness.

The first rival on the scene was Bell Telephone Laboratories, the research arm since 1925 of the sprawling American Telephone and Telegraph Corporation. Bell's early prominence in digital computing has been hidden by the corporation's later wariness about antitrust. The federal government's scrutiny of its nationwide telephone monopoly, especially after a pivotal consent decree in 1956, barred Bell from the commercial computer or electronics markets. In the public view, at least, Bell's own substantial efforts in computing were from the 1950s directed exclusively toward its military systems, especially its high-profile work on the Nike family of missile-guidance systems, as well as its own efforts to automate telephone switching. In effect, Bell built a huge network of sophisticated stored-program computers to automate the telephone system—known as the ESS series and first installed in 1965—but it never called them "computers" nor ever made them available for commercial sale. Beginning in the 1970s, Bell published and widely distributed a series of thick volumes on its historical achievements—at least those that it wished the public to know. The seven volumes of *A History of Engineering and Science in the Bell System* (1975 et seq.) cover all imaginable technical topics—including physical sciences, communication sciences, electronics technology, transmission technology, switching technology, military systems—save one: computing. Nonetheless, despite its efforts to downplay the fact, Bell had an early and sustained pioneering role in computing.[15]

Not only did Bell Laboratories employ the star researchers Claude Shannon and Richard Hamming, who along with Alan Turing created information theory in the 1940s and 1950s, but it also supported George Stibitz's efforts to build electromechanical computers. Shannon, in his 1937 MIT master's thesis, had shown the logical equivalence of networks of telephone relays with statements in Boolean algebra. Bell's telephone engineers used Boolean algebra to analyze telephone networks. Conversely, Stibitz saw that networks of telephone relays could solve problems in Boolean algebra, including the arithmetic of complex numbers, that is, those having both "real" and "imaginary" components based on the square root of minus one. (Complex numbers were not an obscure problem in theoretical mathematics, because telephone engineers used them in designing filters and amplifiers.) He built a Complex Number Calculator using more than four hundred telephone relays for logic and ten telephone crossbar switches for memory, and demonstrated it at a mathematics meeting at Dartmouth College in September 1940. The machine, connected to Dartmouth via a teletype terminal, never left the Bell Laboratories building in lower Manhattan. Bell filed at least five patents on Stibitz's computing inventions between April 1941 and September 1945.[16]

So important was this line of work that Bell actively pursued it well after Stibitz left the company in 1945. During the war, Stibitz experimented with using loops of paper tapes to create simple programs for his calculators, which were utilized for ballistics, antiaircraft, and error-detection computations. In 1946 and 1947, Bell built two copies of Stibitz's so-called model V design. These were immense ENIAC-scale machines with each of them weighing ten tons and consisting of 8,900 telephone relays and costing five hundred thousand dollars. (They were delivered to the National Advisory Committee on Aeronautics, NASA's predecessor agency, and the army's Ballistic Research Laboratory, while Bell built a modified third copy for its own use.) These were fully programmable computers, complete with built-in floating-point arithmetic, internal error checking, conditional branching, and even what might be labeled today as an early two-core central processing unit or CPU. The twin arithmetic units, each with its own set of fifteen memory registers and independent input–output, could work separately on smaller computations or work together in tackling larger ones. In addition to the five Stibitz patents, Bell filed at least twelve further patents on varied computing inventions from 1944 to 1952. So, with good reason, Bell saw its computing work having substantial technical precedence over ENIAC, and consequently filed patent interferences that bottled up the ENIAC patent for years.[17]

The second major rival on the scene was IBM, which got tangled up with Sperry Rand and the ENIAC patent in several ways. The year 1956 was a pivotal one, with antitrust consent decrees issued against IBM as well as Bell. IBM had filed one of the eleven interferences against the ENIAC patent, and IBM and Sperry Rand subsequently filed separate patent infringement suits against each other. In August 1956 the two companies settled their differences with a complex agreement, exchanging patent licenses between them and proposing that IBM pay Sperry Rand ENIAC-based royalties of ten million dollars over an eight-year period. This agreement set the clock ticking, because if there were no valid ENIAC patent issued by 1964 there would be no ten million-dollar payment. (This IBM–Sperry Rand technical exchange was critically scrutinized for its antitrust implications in the final *Honeywell v. Sperry Rand* decision.) Meanwhile, the ENIAC patent with its 148 claims was still proceeding—claim by claim—through a dense thicket of patent interference contests. In September 1961, Bell and Sperry Rand made a similar agreement to exchange patent licenses and technical information, removing Bell's long-running interferences from the docket,

FIGURE 5.10. Bell relay computer Model V, circa 1947. Bell Laboratories built—and patented—a series of sophisticated relay-based computers between 1940 and 1947. Here is the room-sized installation of a fully programmable Model V. Bell's patent interferences bottled up the ENIAC patent for years.

and a supportive ruling a year later by the Patent Office awarded priority to the ENIAC patent, clearing the way for it to be officially issued on February 4, 1964.[18]

Honeywell, itself somewhat of a latecomer to the computing industry, was also a latecomer to the wild and wooly world of patent interferences. Sperry Rand, after collecting its payment from IBM, also went after all the other large computer companies seeking a 1 percent royalty on computing sales. "We were not trying to be greedy," Sperry Rand's patent lawyer later claimed. Sperry Rand opened lengthy and protracted negotiations with GE, RCA, and Burroughs. Ghosts from bygone days reappeared on the scene when members of the original ENIAC team (then represented by Technitrol Company) pressed their claim to be considered coinventors of the ENIAC patent as well as when lawyers from Iowa State University pressed the claim that its faculty member John Atanasoff, a physics colleague of Mauchly's, should also be considered joint inventor—both claims that the Patent Office had earlier denied but that Honeywell successfully resurrected in the main lawsuit.[19]

While other companies met Sperry Rand's challenge variously by strategic "stonewalling" or aggressive counter patent-interference claims, Honeywell began preparations for a full-blown lawsuit against Sperry Rand. Getting wind of this, Sperry Rand on May 26, 1967, filed a standard patent infringement lawsuit against Honeywell in the U.S. District Court in Washington, D.C. (The suit was technically brought by Illinois Scientific Developments, a subsidiary Sperry Rand had formed for the sole purpose of holding the ENIAC patent and, effectively, insulating the larger company from unfavorable claims.) Remarkably enough—on the very same day, and just minutes earlier—Honeywell filed a lawsuit against Sperry Rand in the U.S. District Court in Minnesota. Honeywell not only charged that Sperry Rand had violated the Sherman Antitrust Act (for its cozy 1956 agreement with IBM) but it also sought to invalidate the entire ENIAC patent and so torpedo Sperry Rand's expansive royalty-collecting campaign. Court watchers at the time observed that the District of Columbia court generally was more favorable to patent claims, and so might favor Sperry Rand's side, while Honeywell, as the state's largest private-sector employer, was presumed to have a home court advantage in Minnesota. One year after the remarkable joint filings, Judge John Sirica ruled that the two cases should be combined and heard in Minnesota, noting Honeywell's slim but significant time advantage and arguing that "the convenience of parties and witnesses" and "the interest of justice" would thereby be best served.[20]

From a historian's vantage, what is most important and enduring about the legal enterprise over the next six years took place during the discovery phase as well as in the valuable "exhibits" officially added to the court record during the trial phase. Honeywell submitted 27,259 exhibits to the court record with Sperry Rand adding an additional 7,167 to create a unique historical record. Honeywell's exhibit number one is Charles Babbage's autobiographical *Passages from the Life of a Philosopher* (1864), while its other exhibits aimed to undermine the ENIAC patent and to demonstrate Sperry Rand's alleged violation of antitrust. Among the historical gems preserved in the court record were the top-secret wartime work of Joseph Desch and Robert Mumma at National Cash Register (see chapter 2), little known for years. Likewise, Sperry Rand presented a highly selective set of exhibits, beginning with the entire U.S. Patent Office file on ENIAC, as well as extensive documentation of the early years of the Eckert Mauchly company. The lawsuit, at least, preserved rare and otherwise unavailable documentation on the early years of computing. An immense "master record" consisting of 211 rolls of microfilm combined the complementary records held at the Charles Babbage Institute, the Hagley Museum and Library, and the University of Pennsylvania's archives. Records from the presiding judge and attorneys round out a very large historical corpus.[21]

Most valuable to historians is the set of detailed on-the-record depositions taken from eighty well-known figures in the history of computing. Lawyers from both sides pointed sharp questions at key figures in the ENIAC effort (including Herman Goldstine, Frances Holberton, John Holberton, Ruth Teitelbaum, Jean Bartik, and especially John Mauchly, who was run through the wringer), as well as many other leading computer figures, such as Joseph Desch, Jan Rajchman, Warren Weaver, Vladimir Zworykin, and last, but by no means least, John Atanasoff. Posttrial assessments of the depositions have been polarized and full of controversy, paralleling the sharply divergent views on the trial itself. Latter-day proponents of John Atanasoff in supporting the judge's decision have, not surprisingly, rather harshly scrutinized Mauchly's deposition. They frequently mention Mauchly's reticence in giving the deposition, often commenting on his allegedly diffident, vague, or unhelpful answers. In my own reading of his deposition, I found Mauchly understandably struggling to pin down events from thirty years or more in his past, including where he had gone to school as a child, the street numbers of the several houses where his family had lived, and the courses required of freshmen at Johns Hopkins University; the two rival sets of lawyers certainly jousted over which questions posed to Mauchly were proper and relevant. I also found in his deposition clear references to a significant but "classified" product that Eckert and Mauchly delivered to the National Security Agency (see chapter 1) that is otherwise unknown and, it seems, entirely undocumented.[22]

Judge Earl Larson's final decision, like nearly everything else about *Honeywell v. Sperry Rand,* defies easy description. Larson was a Minnesota native, University of Minnesota alumnus, and trial attorney who had been named by President John F. Kennedy to the U.S. District Circuit for Minnesota. "Patent cases are difficult for judges," he wrote later, noting that he had tried to engage an outside expert in electrical engineering and law (a former University of Minnesota law professor), but when the expert declined the assignment, "I had to do it on my own."[23] Handing down his decision on October 19, 1973, seven months after sitting through nearly two years of court proceedings, Larson set down his findings in a complex 248-page document. It was obvious to anyone paying attention that the ENIAC patent was in serious trouble. Larson nonetheless ruled that Eckert and Mauchly were the sole inventors of the ENIAC, setting aside the hopes and claims of half a dozen potential coinventors and, it may be suggested, creating animosities that persist until today.

One of the individuals so slighted by Larson was Arthur Burks. A Minnesota native, born in Duluth, Arthur Burks (according to the court's findings) "made major contributions to the design of the accumulator and multiplier of ENIAC and signed at least 77 drawings" while at the Moore School, then joined John von Neumann's computer project at the Institute for Advanced Study in Princeton (1946) before spending his career as a computer-science faculty member at the University of Michigan. Burks, along with his wife Alice (the two met at the Moore School), have for three decades since the trial been among the most severe critics of John Mauchly and among the staunchest supporters of Atanasoff. While readily acknowledging the "work, experiments, and suggestions" of Burks and others, Larson stated that because patents in all respects are presumed to be valid, "Honeywell has a heavy burden," which it failed to meet, to prove that the others' "work, experiments, and suggestions" rose to the level of inventiveness.[24]

Although it lost this skirmish, Honeywell clearly won the wider war against the ENIAC patent. The most serious blow was that Larson ruled it outright "invalid" owing principally to its tardy filing back in June 1947. This was not really news. On the letter of patent law, Eckert and Mauchly had simply failed to file their patent application within one year of the first public disclosure of the invention, whether one considered this to be the circulation of John von Neumann's "draft" report on the ENIAC's successor (June 1945), Eckert and Mauchly's own official report on ENIAC's successor (September 1945), or the ENIAC's high-profile public dedication (February 1946). All these occurred *before* the patent application's "critical date" of June 1946. The ENIAC patent was also ruled "unenforceable" on a number of other serious accounts, including a number of its many claims being anticipated by prior patents, as well as irregularities in the late modifications by the Sperry Rand patent attorneys.

It is little regarded today but Larson's lengthy finding number 15 might have ignited a round of antitrust proceedings in the computer industry. As noted earlier, IBM was already on the federal government's antitrust watch list owing to the 1956 consent decree. Perhaps harking back to his wartime service in the Office of Price Administration, he found stern antitrust implications in the 1956 agreement between IBM and Sperry Rand. Larson found the two companies' patent cross-licensing agreement—which involved sharing technical resources between the two companies, giving advantages to each of them exclusive of any other computer companies—to be unfair restraint of trade and thus a violation of the Sherman Antitrust Act. Although only IBM was successful in this attempt at monopoly, thought Larson, both companies were equally guilty under the law. All the same, despite the seeming gravity of these matters, he imposed no penalty on Sperry Rand. Recall that Honeywell had specifically charged Sperry Rand with antitrust violations, while IBM was no part of the lawsuit.

For computer history, however, it is Larson's finding 3 that overshadows all else. There, strikingly, he wrote that "Eckert and Mauchly did not themselves first invent the automatic electronic digital computer, but instead derived that subject matter from one Dr. John Vincent Atanasoff."[25] Derivation is a grave offense, because in patent law "derivation will bar the issuance of a patent to the deriver." It suggests an inventor's consciously following a known technical precedent, whether patented or not, and it is not simple to establish, because "derivation requires complete conception by another and communication of that conception by any means to the party charged with derivation."[26] Not every conflicting patent case involves derivation. For instance, while certain of the ENIAC patent's lesser claims resembled a little-known 1942 patent of IBM, and indeed were found to have been anticipated by it, those claims were merely rendered "invalid" because Eckert and Mauchly did not know about the IBM patent. This instance did not involve "complete conception" and "communication of that conception."

For Judge Larson, a crucial point was Mauchly's 1941 visit with Atanasoff. Larson maintained that because Mauchly in June 1941 had spent four days in Iowa with Atanasoff discussing physics and computing, Mauchly consequentially "derived" the design for ENIAC from his work. Mauchly and Atanasoff were physics colleagues at the time, as well as later, and Atanasoff had invited Mauchly to Iowa six months earlier when he was visiting Mauchly in Philadelphia while on a trip to do patent research in Washington, D.C. The historical evidence is frustratingly incomplete about just what Mauchly saw, or might have seen, in Atanasoff's lab in Iowa, and the rancorous lawsuit and its aftermath have contaminated everyone's sense of fair play, considered judgment, and good humor.

In effect, Judge Larson found Atanasoff's version of the visit more credible than Mauchly's. In his version of events, Mauchly recalled that Atanasoff evinced a "conscious deliberate hesitation" to discuss all aspects of the computer, owing to Atanasoff's own patent concerns, and that he had seen only partially complete results and "not yet an operable machine." He understood there was "a kind of binary adder" but that "the precise circuits whereby this was accomplished were not communicated to me." A key liability of the Atanasoff machine, "a poor way to start out," as he reported to Eckert when back in Philadelphia, was its memory scheme. This involved brushes, amplifiers, and "mechanical commutation" needed to refresh the electric charges used to store numbers.[27]

Judge Larson's official court findings depict a different visit in Iowa and certainly describe a different machine. The Atanasoff machine, in the version of sharply conflicting stories that Larson codified in his decision, had evolved from "an operating breadboard machine" existing already in 1939 to a full working "prototype or pilot model." It was, if not "entirely complete" at the time of Mauchly's 1941 visit, "sufficiently well advanced so that the principles of its operation, including detail design features, was explained and demonstrated to Mauchly" in a "free and open" manner with "no significant information . . . withheld." Larson's finding pivoted on his belief that Mauchly, while "broadly interested in electronic analog calculating devices," such as his own harmonic analyzer, prior to his Iowa visit "had not conceived an automatic electronic digital computer."[28]

There is simply no way to square these rival accounts of the Iowa visit, because here Larson entirely set aside Mauchly's earlier and substantial experiences with electronic digital computing. These experiences, as the court evidence itself makes clear,

FIGURE 5.11. *Honeywell v. Sperry Rand* exhibit of Atanasoff computer, circa 1940. The computer was abandoned when Atanasoff departed Iowa State University for wartime work in 1942. This photograph, introduced as an official court exhibit, helped persuade Judge Larson to void the ENIAC patent.

included the counting and scaling circuits common in cosmic-ray physics since the 1930s, Mauchly's having seen Stibitz's relay computer in action at the September 1940 meeting in Dartmouth ("this was a very interesting demonstration," he noted), and even Mauchly's own notes and diagrams on "computing circuits" of some sophistication—containing up to 150 vacuum tubes—dated January and April 1941, well before his Iowa visit.[29]

Atanasoff's supporters, including a recent book by novelist Jane Smiley, never tire of quoting one of Mauchly's letters, seeming to inquire about a go-ahead "to build an 'Atanasoff Calculator' (a la Bush analyzer) here" at the Moore School. The letter, especially when only this snippet is quoted, appears to be strong stuff. Yet, if Mauchly ever actually saw the entire "Atanasoff Calculator," sometimes dubbed the ABC machine in recognition of the graduate student Clifford Berry who assisted Atanasoff, he somehow did not learn the most important lesson that he might have. The

Atanasoff computer was at its core a thoroughgoing binary-based calculating engine. It should be properly remembered and celebrated for featuring an electromechanical unit that converted base-10 decimal numbers into base-2 binary ones (up to fifteen decimal places could be represented in fifty binary bits). ABC then did arithmetic computations entirely with logic circuits built with three hundred vacuum tubes, and finally returned the binary results to the electromechanical binary-decimal converter to regain the decimal-based results. Atanasoff had designed his machine to solve systems of linear equations, with up to twenty-nine equations and twenty-nine variables. A video made with the 1997 reconstruction of the ABC machine makes clear that substantial human input was required, at each step of a calculation, to set switches and move cards in and out of the machine.[30] It was never designed to be anything like the general-purpose ENIAC, but if only he had grasped the compelling logic of binary math, Mauchly might have cut ENIAC's immense number of troublesome vacuum tubes by as much as two-thirds and dispensed with ENIAC's convoluted "ring-counting" scheme. Mauchly was a smart man, but somehow in Iowa he missed the main lesson.

Honeywell v. Sperry Rand is a multifaceted monument in computer history, firmly rooted in Minnesota. It was the largest such federal lawsuit at the time, and it prominently featured Minnesota companies as well as a Minnesota courtroom and a Minnesota judge. It is fitting that the voluminous *Honeywell v. Sperry Rand* records became the very first archival collection of the Charles Babbage Institute, donated by Honeywell itself in 1984. The legacy of this lawsuit persists. It once seemed merely quaint that so many people were so vitally interested in establishing the "real" priority in inventing the computer, for it appeared to be largely a question of self-interested insiders debating the fine points. Yet, as computing in more recent years has emerged as a major economic and cultural force, the cultural salience of computing's origin moment seems actually to be growing. The debates on ENIAC and its rivals continue to be strongly argued in the popular media, even if the instrumental importance of the ENIAC patent has long since faded.

Honeywell Project (1968–90)

It is not only Honeywell's technical achievements and legal strategizing that make it likely, in the long run, to be the most famous of all the Minnesota computing companies. Honeywell's national prominence as a military contractor attracted some unusual local attention. The Honeywell Project was a long-lived antimilitary group that staged regular protests at Honeywell's Minneapolis headquarters and elsewhere in the Twin Cities for more than two decades (1968–90). It took form in the ferment of the 1960s after Marv Davidov, a seasoned activist from the civil rights and draft resistance movements, considered Honeywell's high-profile participation in the Vietnam War. "They were number ten on the list of national war contractors . . . [making]

cluster bombs . . . guidance systems for nukes, first strike nuclear weapons," as he put the case against Honeywell some years later. In 1969, Davidov and two sympathetic faculty colleagues from the University of Minnesota, mechanical engineer Edward Anderson and physicist Woods Halley, met in person for two hours with Honeywell's CEO James Binger, who told them that it was not only in North Vietnam where the Honeywell-manufactured cluster bombs were deployed but also, he admitted, in South Vietnam. Davidov's group pressed Binger to consider "conversion" from producing weapons such as fragmentation bombs, land mines, and napalm to the making of civilian goods, offering to bring in Colombia University industrial engineer Seymour Melman as a consultant, but the conversion proposal was not actively taken up by the corporation.[31]

In an era when antiwar protests could end violently, as they did at Kent State or in the Chicago police riot, the Honeywell Project's legacy of vigorous yet peaceable protest is a true Minnesota story. Notably, the Minnesota Historical Society recently

FIGURE 5.12. Honeywell security officer Fred Carey *(left)* discusses ground rules with Honeywell Project marchers including Marv Davidov *(right)* on December 12, 1969. Photograph by Jack Gillis.

named the Honeywell Project as one of the 150 events that shaped the state's history, alongside such familiar state icons as Bob Dylan, Paul Bunyan, the Minnesota State Fair, and Spam.[32] A distinctive local culture supported the protest movement. Davidov recalled that information flowed along "a conveyor belt" via the University of Minnesota's science and engineering faculty from "Honeywell Ordnance where we got information . . . people on the inside would ship us stuff, send us stuff as we were leafleting and keeping it up because endurance is the key." The persisting protests at Honeywell's South Minneapolis corporate headquarters, its suburban factories, and even chairman Binger's Episcopal church were unusually civil in their conduct, since agreed-upon ground rules had been negotiated beforehand with police and corporate officials. One Honeywell security officer even offered the protesters his company's protection "if hotheads in the [St. Louis Park] plant pull any funny stuff."[33]

Honeywell was accommodating to the protesters in part because several leading lights of Minneapolis society—including Charlie Pillsbury and Mark Dayton, respectively, great-grandson of the flour founder and scion of the department store family—publicly sided with the efforts to alter the company's military strategy. Pillsbury sought to use his ownership of a hundred shares of Honeywell stock to gain an inside voice at the company's annual shareholders meeting, with some success. "I participated with Marv Davidov in the very first demonstration of the Honeywell Project, at Honeywell's annual meeting in 1970, and voted to remove my father from the board of directors of Honeywell," Dayton stated some years later, in between his serving in the U.S. Senate (2001–7) and becoming governor of the state in 2011.[34] Such national figures as activist Jerry Rubin, antiwar poet Robert Bly, and others regularly attended Honeywell Project rallies and addressed the faithful.

Not everything about the Honeywell Project was sweetness and light. The antimilitary protesters faced open hostility from the pro-military Teamster's Union employed at the company's factories and, it turned out, secret harassment from covert FBI infiltration and provocations. A huge march on Honeywell's headquarters in spring 1970 ended badly when some protesters, "perhaps including paid informants, broke their agreement . . . and pushed the [company's] door in. The demonstration became a riot of broken windows and mace." The FBI amassed a 1,200-page file on Davidov, and later settled a federal lawsuit by paying seventy thousand dollars to the Honeywell Project. The Honeywell Project revived with the antiwar protests accompanying the Reagan-era military buildup. The mass arrest in April 1983 of perhaps 150 protesters, one of whom happened to be the Minneapolis police chief's wife, Erica Bouza, prompted widespread national media attention, and she and her husband become an odd specie of celebrities. As police chief, Tony Bouza, accustomed to the rough-and-tumble of New York City, commented: "This is very civilized, I am used to being denounced as fascist swine. I guess I still am not used to the Midwest way of doing things."[35] In the Midwest way, the police chief repeatedly drove his protester-spouse

FIGURE 5.13. Honeywell's civilian-sector "literate laser" for data storage, 1967. Honeywell worked on military lasers for years, then sought publicity for its Minneapolis-based research in 1967, even before the Honeywell Project sought to convert the company to civilian markets. The "literate laser" was an early form of CDs and DVDs.

to Honeywell Project events, never quite certain if she would be coming home on her own in a taxi or with arresting police officers in a squad car.

Local protest activities, while shifting focus over the years, continued in the Minneapolis suburbs even after Honeywell reorganized its military business as Alliant Techsystems in 1990. Indeed, the local activities known as AlliantAction ceased only in October 2011 when Alliant's corporate headquarters were relocated to the Washington, D.C., area. Alliant, at the time "the largest supplier of rifle rounds to the U.S. military," made the move "to be closer to the Defense Department and the U.S. Congress."[36]

6
Big Blue

Manufacturing and Innovation
at IBM Rochester

IBM was the first computer company to consciously choose Minnesota as a base of operations. The state's other computing ventures had arrived somewhat by happenstance or grown organically as start-ups or spin-offs. Back in 1946, the location of the Engineering Research Associates was largely a matter of luck, when the navy intelligence officers in Washington, D.C., and Dayton, Ohio, recruited John Parker and his empty glider factory in St. Paul to the cause. The expansion of ERA's corporate parent Sperry Rand Univac into a major computer employer in the state resulted from building incrementally outward from the ERA core. Similarly, no one seriously considered setting up Control Data anywhere besides the Twin Cities because one of its chief assets was, indeed, the many Univac engineers and managers it came to employ itself. Even Honeywell, creating a loosely focused national network of computer efforts, had its roots firmly in Minneapolis for more than a century. But the arrival of IBM in Rochester, Minnesota, signaled a new epoch for the state.

Beginning in 1956, IBM connected Minnesota to the largest computer company in the world. Originally selected as a new "greenfield" factory site, IBM Rochester in subsequent years built a stunning variety of tabulating machines and computing systems. It also created a development laboratory that assisted with the design, engineering, and development of an array of products and was itself awarded 2,700 patents over three decades, adding an additional 440 patents in 2009 and 500 in 2010.[1] The range of Rochester-developed and -manufactured products includes, most notably, IBM's "mid-range" computing systems, a line of high-performance disk drives, the Blue Gene supercomputers in recent years, and a generation of superfast gaming chips today. As a division in a giant global corporation, Rochester has contributed to many joint IBM developments efforts, collaborating with researchers in Yorktown Heights,

New York, and San Jose, California, and other facilities. It may sound odd, but for much of its history IBM Rochester functioned very much like an industrial district unto itself, combining (often in informally structured ways) the strengths of multiple divisions and specialized skills. Sometimes, it did so because of explicit instructions from IBM headquarters; other times, it did so despite any such instruction.[2] In at least one instance, Rochester successfully launched a major innovation with a significant degree of stealth.[3]

Viewed from the local perspective in Rochester, IBM looms large with its four hundred-acre site on the city's periphery, its peak employment of 8,100 people in the early 1990s, and most of all its sprawling 3.6-million square-foot building.[4] The building is, as of this writing, still the largest IBM facility under one roof in the world. Yet, from the wider corporate perspective, IBM Rochester is a small cog in an immense global enterprise. Rochester's deep strengths in manufacturing, designing, and developing a wide range of IBM products—from the electromechanical tabulators of the 1950s through to today's supercomputers and gaming chips—have contributed handsomely to IBM's bottom line. In 1990, IBM's middle-range computer division, centered on IBM Rochester, had greater revenues than the entire Digital Equipment Corporation, the number two computer company in the United States. In financial terms, one could say that IBM Rochester was the second-largest computer company in the world.[5] Yet IBM, like most American corporations, has over the years systematically downplayed its own achievements in manufacturing and instead focused its corporate publicity on "research-and-development" (R&D) efforts with perceived greater prestige.

While IBM profited from Rochester's high-volume manufacturing, its corporate publicity gave the spotlight to the company's R&D laboratories in places like Yorktown Heights. To give just one example, IBM's corporate Web site portrays its Blue Gene supercomputer to be a triumph of efforts by IBM Research, which are impressive enough, while entirely skipping over the development, design, and manufacturing of Blue Gene done at Rochester. This chapter tells the untold story of how Rochester transformed a research prototype into the world's fastest supercomputer installed at Lawrence Livermore National Laboratory. Similarly, IBM's recent "Watson," its *Jeopardy!*-champion supercomputer, is hailed as a triumph of IBM Research—with no mention that it ran on "10 refrigerator-sized racks filled with 90 Power 750 servers built in Rochester."[6] This imbalance is not likely to improve. Since the mid-1990s, IBM's corporate leadership has embraced "information technology services" as the key to its future, tending to further marginalize computer hardware development and manufacturing. This chapter sets IBM Rochester within the context of a global corporation with a long-term history and complex culture.

Birth of Big Blue

IBM traces its origins to an idiosyncratic merger that formed the Computing Tabulating and Recording Company in 1911. Oddly enough, neither of the company's two principal early figures—Herman Hollerith and Thomas J. Watson Sr.—were leaders at the time of the merger. The technical centerpiece of C-T-R was the brainchild of Herman Hollerith, an inventor and engineer from Buffalo, New York, who transformed a basket of wires, wood, and metal into the basis of the tabulating machine industry. Hollerith's machines, developed initially for the U.S. Census Bureau, responded to the rising tide of information generated by railroads, insurance companies, banks, and the U.S. federal government. Hollerith won a competition for the contract to process the 1890 census. At the time, no one knew how many people there were in the expanding county—the most expansive estimates hoped for seventy-five million—let alone how many people lived in cities, towns, or farms and how many lived in each state, a crucial number for the apportionment of taxes and political representation. Every ten years, a small army of census workers fanned out across the country and filled out sheets of paper enumerating the population, agriculture, manufacturing, and other measures of the country. Prior to Hollerith, a second small army of workers in Washington, D.C., painstakingly summed up the mountain of census forms to form a statistical portrait of the country. The 1880 census took an agonizing seven years to complete. For an impatient country, the wait seemed unbearable. For the 1890 census, Hollerith transferred the census data onto rectangular stiff paper cards with twelve rows and twenty columns, with punched holes recording the data. Hollerith's punches, tabulators, and sorting machines mechanically prepared the national statistics. Within six weeks he announced the U.S. population was nearly sixty-three million. The census was complete in one year.

While Hollerith gained a Ph.D. at Columbia University for "An Electric Tabulating System" (1889) describing his system, he also filed for a series of patents and created the Tabulating Machine Company in 1896. That same year, Thomas J. Watson, a struggling young salesman in Buffalo, New York, joined the National Cash Register Company. NCR was the personal fiefdom of John Patterson. Patterson created a scientific method of salesmanship—staging the salesman's "approach," "proposition," "demonstration," and "close," programmatically set down in the NCR *Primer,* and later in the *Book of Arguments* and two hundred-page *Sales Manual,* establishing clear sales incentives, and offering ample sales rewards including the "100 Percent Club"—that was taken up by many companies over the years.[7] He cultivated a strong sense of corporate culture. "THINK," Watson's pervasive IBM slogan, even had its origin in his NCR days. Infamously, however, Patterson also high-handedly fired his staff (including Watson) for apparently inconsequential violations. He landed afoul of federal law when his executives practiced strong-arm business practices. Watson for years ran a shadow company funded by NCR that bought up competitors' old cash registers, monopolizing that market and clearing the way for an NCR salesman to sell new ones. At the time of the C-T-R merger in 1911, Watson was still with NCR, by then its national sales manager; in the next two years, he would be indicted for antitrust violations, along with thirty others, including Patterson, and then summarily fired by Patterson.

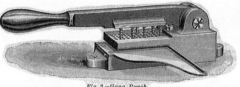

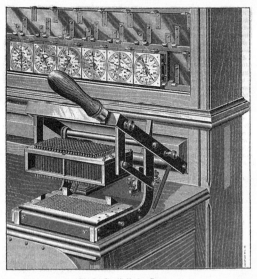

FIGURE 6.1. Hollerith tabulating machines, 1895. (a) Pantograph card punch *(top)* and sorting machine, with cutaway showing card boxes; (b) Gang punch *(top)* and circuit-closing press (with tabulating counters in background). Source: "Hollerith's Electric Tabulating Machine," *Railroad Gazette* (April 19, 1895): images 4–5. Library of Congress.

In 1911, Hollerith, suffering from ill health and a foul temper, was only too happy to sell his company to Charles Flint, a serial business promoter who had earlier completed mergers in rubber, chewing gum, and woolens. These mergers have been largely forgotten. Flint, later that year, created C-T-R by merging together Hollerith's tabulating company with a meat-scale company from Dayton, Ohio, and two time-clock companies from upstate New York, which eventually gave rise to the IBM manufacturing district centered on Endicott. Hollerith served C-T-R as a consultant but did not play any constructive role in running the company. In April 1914, Watson convinced Flint to hire him as general manager at a salary of twenty-five thousand dollars a year, 5 percent of C-T-R's net profit, and 1,200 shares of stock; a year later, cleared of the NCR antitrust violation, Watson became president. He brought to the new venture Patterson's sales methods and corporate culture, as well as his own fierce determination to succeed. By 1924, Watson signaled his ambitions when he renamed C-T-R, rather grandly, as "International Business Machines."

FIGURE 6.2. IBM corporate logos from 1911 to 1956 hint at the transformations of the company's image. The C-T-R logo in 1911 was a traditional design. The "world" graphic in 1925 signified the global ambitions of the newly named International Business Machines. In 1956 IBM embraced modernism and industrial design. Source: IBM.

From the 1920s to the 1940s, IBM grew from being the smallest of the four "office machine giants" to the largest and most profitable. Hollerith's valuable patents were a start, and the company expanded its time-clock factories in Endicott, New York, into a major manufacturing complex. In 1928, IBM's revenue and profit figures placed it fourth behind the leaders Remington Rand, National Cash Register, and Burroughs. Difficult times followed the stock market crash of 1929 and the Great Depression. Watson determined to keep his factories running, piling up tabulating and accounting machines in the company's warehouses. It was a risky gambit, and many thought it outright folly—until the moment Franklin D. Roosevelt signed the Social Security Act of 1935. "No single flourish of a pen had ever created such a gigantic information processing problem," writes Watson's biographer.[8]

At once, the U.S. federal government created the need for a permanent and cumulative record on every single person in the workforce (perhaps twenty-five million). In addition, for the first time private-sector employers were required to keep permanent records of their entire workforce, requiring even more tabulating machines. (Income taxes at the time were paid by only 5 percent of U.S. workers.) Not many firms besides IBM had a warehouse full of machines capable of doing the task. Besides its tabulating machines, Endicott was shipping between five and ten million cards a day in boxes packed with ten thousand cards. By 1939, IBM's annual revenues at $40 million were on a par with its leading competitors and its profits at $9.1 million were between three and fives times larger. During World War II, IBM opened a new factory at Poughkeepsie, New York, and further expanded Endicott, selling thousands of punches, sorters, and tabulating machines to the military services. With the expansion of government and business after the war, IBM was annually producing sixteen *billion* punch cards by 1952.

IBM launched an impressive effort in electronic digital computing in the 1950s, even if it is difficult to pin down the precise year it did so. Clearly, along with the rest of the world, IBM was negotiating a sea change in information processing. By most accounts, Thomas Watson Sr. was wary of the new field of computing. IBM's early experiments in the field were glorified electromechanical calculating machines, such as its Automatic Sequence Controlled Calculator (1944), also known as the Harvard Mark I (see chapter 1), or the Selective Sequence Electronic Calculator (1948) that the company built entirely on its own. At Watson's direction, the IBM 701 (1953) was labeled as the Defense Calculator, although with its memory and calculating circuits and stored program it certainly was a state-of-the-art digital computer. There is no evidence that he ever said, even though he is frequently so quoted, that the world needs only five computers.[9]

By the mid-1950s, IBM was developing a range of upper-end computers as follow-ons to the 701 as well as the lower-end IBM 650, the rotating magnetic-drum machine (which is connected to St. Paul's Engineering Research Associates, as chapter 2 related) that sold a phenomenal two thousand copies beginning in 1954.[10] An immense shot in the arm came in 1954 when IBM landed the production contract for the SAGE computers, with five hundred million dollars in revenues coming to IBM for building fifty-two of the vacuum-tube behemoths as well as the invaluable exposure to the company of MIT-based computing and networking technology.[11] By 1956, when Watson Senior turned over leadership of IBM to his son, Thomas J. Watson Jr., the company with fifty-six thousand employees had its feet firmly planted in both the established tabulating industry and the developing computer industry. Sales of punched cards alone still made up 30 percent of the corporation's profits.

IBM's decision to create a new manufacturing facility made sense only in the context of the vibrant and expanding punch-card industry. IBM examined eighty Midwestern cities before naming Madison, Wisconsin, and Rochester, Minnesota, as finalists. Originally a campground for wagon-train settlers in the 1850s, Rochester was named for the upstate New York town, and it grew slowly as the seat of Olmsted County in southern Minnesota. Except for the activities of the Mayo medical family in the 1880s, "Rochester would have undoubtedly developed into a pleasant little trade and railroad center for the farmers whose rich claims surrounded it," according to the 1938 *WPA Guide to Minnesota*. As it happened, the Mayo brothers—William and Charles (who both died in 1939)—built the Mayo Clinic, founded in 1889 on the base of their father's flourishing medical practice, into a specialty clinic with a national and

FIGURE 6.3. IBM built its new facility on the outskirts of Rochester, Minnesota, in 1956, where previously farmers' fields had existed for several generations. Source: IBM.

soon enough worldwide reputation. Already by the 1930s up to 150,000 visitors seeking medical treatment arrived each year, from the four corners of the world, lending to Rochester "a cosmopolitan air wholly unexpected in a southern Minnesota city." A nearby private airport connected Rochester with the world. The only industrial enterprise noted by the WPA was "a canning factory topped with a huge ear of corn."[12]

In IBM's studied assessment, Madison and Rochester each scored well in such categories as schools, utilities, taxes, and transportation as well as in the more elusive category of "morals," but while Madison had a larger labor market it also had labor unions. Rochester at the time, since the wartime industrial mobilization had largely passed it by, had still just one canning factory with seasonal employment of several hundred. In February 1956, IBM announced that it would purchase 397 acres of farmland two miles from the Rochester city center for a new permanent facility, begining operations in leased space later that summer with 174 workers. Workers began assembly immediately, and within two months they shipped two model 077 numeric collators to customers in Iowa and Texas. Rochester's early products were straight from the punch-card tabulating world, including numeric collators, reproducing punches, ticket converters, and the input/output unit for the IBM 305 RAMAC disk drive. Selected Rochester employees travelled to Endicott, IBM's mammoth factory complex in New York, where they spent nine months learning to make new models of several types of punches.

FIGURE 6.4. IBM Rochester first gained fame for its modern and efficient assembly line (here circa 1960) as well as its development of new lines of computers. Source: IBM.

Rochester was much more than a mere factory site to IBM. Rochester's factory complex was one of the assertively "modern" buildings that IBM commissioned under the new president Thomas Watson Jr. In putting his stamp on the company, Watson hired Eliot Noyes in 1956 and gave him the grand title of Consultant Design Director with a company-wide scope. Noyes had been converted to the modernist cause at the Harvard Graduate School of Design under the influence of Walter Gropius, Marcel Breuer, and Le Corbusier. At the Museum of Modern Art as a curator (1940–46)—he met Watson at the Pentagon during the war years—and subsequently in his architecture and design practice, Noyes aimed to bring the aesthetic and technological vision of modernism into American life. "One thing I am not going to become is a guy who is called in to change the expression on the corporate face by hanging abstract paintings on the office walls," he stated.[13]

Working closely with IBM over two decades, Noyes created such modern icons as the IBM Selectric typewriter, originally conceived as a utilitarian printing device. He also hired an A-list of creative talent, including designers Charles and Ray Eames, famed Finnish architect Eero Saarinen, graphic artist Paul Rand, and arch-modernist Mies van der Rohe.[14] Although much of the corporate publicity at the time, and architecture history ever after, has focused on the showcase headquarters buildings and research campuses, the numerical majority of the 150 buildings commissioned by IBM (during a building spree that lasted from 1956 to 1971) were, like Rochester, more humble.[15] Saarinen's design for a massive and assertively modern blue-tinted glass building signaled that IBM was in Rochester for the long haul. The Rochester building, commissioned in 1956 and opened in 1958, paved the way for IBM's showcase Thomas J. Watson Research Center in Yorktown Heights, also designed by Saarinen and opened in 1961.

One other matter concerning Rochester's connection to IBM's corporate image deserves comment, and that is the puzzling origin of the phrase "big blue." "IBM has been known informally as Big Blue for decades, but the origin of the colorful nickname remains a mystery," noted one of the commemorations of IBM's recent corporate centennial. If you start counting backwards, it is hard not to arrive at some significance for Saarinen's assertively blue design, which, he suggested, was inspired by the Minnesota sky. "The vibrancy of the two blues, which helps avoid monotony at close view, changes when seen from the distance. Then the total effect is a dark blue band making a transition from the tawny-green, rolling landscape to the sky. In winter, the blue vibrates with greater intensity across the snowscape," he noted. "The result is a building made up logically and appropriately for IBM of precise, machine-manufactured parts."[16] Might there be some relation between his big blue building and the IBM nickname? "Big Blue" was a commonplace for IBM by the 1980s when it routinely appeared in the business pages. Innumerable articles and a small shelf of books by admirers and critics alike played on the nickname (*Big Blue: IBM's Use and*

Abuse of Power or *Big Blues: The Unmaking of IBM* or *Who's Afraid of Big Blue?* or *Saving Big Blue*). A 1981 article in *Business Week* seems to be the first appearance in print: "I don't want to be saying I should have stuck with the 'Big Blue', says one IBM loyalist. The nickname comes from the pervasiveness of IBM's blue computers."[17] Paul Rand's 1972 logo for the corporation, a stylized "IBM" sliced by eight horizontal bars, was often done in blue.[18] And while it's sometimes suggested that IBM switched to a standard blue in the 1960s, its signature System/360 computer actually came "in five standard colors."[19] There are other, speculative suggestions pointing to IBM's conservative dress code or IBM's being a "blue-chip" stock. Not even IBM seems certain about the origin.[20] One thing is clear, however: the Rochester building, from 1958 on, was very blue and became very big. Perhaps there is more to its playful local nickname of "Big Blue Zoo" than meets the eye?

Rochester was involved with product engineering and new product development even before Francis (Dutch) Fairchild was named first manager of the Rochester Development Lab in 1961. The combining in Rochester of engineering, development, and manufacturing constituted something of an industrial district, even if the whole was owned by a single company. Fairchild was an experienced twenty-five-year veteran of the company, with a background as administrative manager of IBM Endicott's development laboratory. Already, three years before his appointment, Rochester engineers had modified existing products coming down the assembly lines to create two new models of card collators. Engineers and managers worked diligently to achieve efficiencies in manufacturing, so that by 1961 they had achieved an "integrated assembly

FIGURE 6.5. Was IBM Rochester's large blue building the source for the company's "Big Blue" nickname? Photograph by Jonathunder (June 18, 2006); reprinted with permission. For a color version of this photograph, see tinyurl.com/5vvhu8f.

line" that trimmed costs by nearly 20 percent through the streamlined handling of parts and materials. The next year, Rochester installed an IBM 1401 computer system, alongside its existing 650 model, for devising assembly schedules and forecasting parts requirements. Under Fairchild, Rochester's development lab interacted closely with the manufacturing activities. Much of its early work involved redesigning machines that were based on vacuum tubes and electromechanical relays instead to employ solid-state electronics and magnetic-core memories. This state-of-the-art technical base permitted Rochester to become IBM's lead manufacturing site for more than two dozen models, including high-speed punches, readers, sorters, and collators as well as the card reader-punch for the IBM 1401 computer system, a breakthrough top-selling transistorized computer. Rochester also took over manufacture of the IBM 803 Proof Machine that clerks used to "sort, list, prove and endorse checks, sales tickets, vouchers and other business documents in a single operation."[21]

Although it is difficult to document the innumerable incremental changes resulting from process engineering, such as an "improved technique of wiring electric panels" described in 1961, we can with some precision identify the Rochester-based development efforts that resulted in new IBM models. While Rochester was initially dependent on products developed at the Endicott or Poughkeepsie manufacturing facilities, it soon developed a significant independent capacity for technical innovation. In the 1960s alone, the Rochester-developed and -manufactured products include these new or redesigned IBM models: the 188 collator, 1060 banking system, 1030 data collection system, 2540 card reader-punch, 1287 optical reader, 2780 data transmission terminal, 1288 optical page reader, and 2502 card reader. The development lab and its associated engineering groups also worked on more general projects, such as "mechanical synthesis" used in electrical power engineering, alphanumerical character recognition, computer analysis of electrocardiograms and X-ray images, and plastic molding techniques. The most significant Rochester development, examined in the next section, initiated a notable line of "mid-range" computers that formed Rochester's bread and butter for two decades.

Especially in our age of global R&D, it is worth drawing attention to the locally situated synergies that resulted from Rochester's model of closely coupled "D&M," where design, development, and manufacturing activities were in close physical proximity and continual productive contact. These are the very hallmarks of an industrial district, such as St. Paul's Midway or South Minneapolis. "The manufacturing arm and the engineering arm worked very closely together," stated Tom Paske, who started at Rochester as a prototyping machinist in the 1960s. "The facilities . . . were joined together with a hallway, so all you had to do was walk down the hallway and you could talk to the person who was in charge of a part. It made it easier for the manufacturing engineers and the development engineers who were working on it." Rochester evolved an integrated model where development and manufacturing engineers worked jointly on a prototype line, actively moving new products from development into manufacturing production.[22] Software developers in Rochester experienced the same. One lesson from Rochester's model of "high-tech" manufacturing is the value of continual interaction between researchers, designers, developers, programmers, and manufacturing experts. The locally situated dynamics of industrial districts can take place within large companies, such as IBM, as well as between smaller or medium-sized ones in urban or suburban locations.

Minicomputing

IBM Rochester's most notable technical and commercial achievement—its series of "mid-range" computers that might elsewhere have been branded as "minicomputers"—occurred in the shadow of IBM's massive System/360 effort. In that oft-dissected chapter in the corporate history, IBM famously "bet the company" on a new line of compatible mainframe computers. *Fortune* magazine breathlessly called it "a $5 billion gamble." IBM customers needing a bit more computing power could trade up within the System/360 family, for the most part running the exact same software on their new IBM machine. IBM's ultimate success with System/360 propelled the company into a unique dominant position in the computer industry, which it maintained from the late 1960s through the early 1990s. Needless to say, IBM's corporate publicity made the most of the System/360 and its direct lines of successors. Little appreciated at the time, at least in the company's public image, was the series of mid-range computers designed, developed, and manufactured at Rochester. The most famous of these, the AS/400, sold a phenomenal four hundred thousand copies, generating annual revenues of $14 billion in 1990. At that time, according to one IBM watcher, "an AS/400 shipped out of IBM's factories every 12 minutes and the AS/400 business represented about 15 percent of IBM's annual revenues."[23] Also, just two years after the launch of AS/400, IBM Rochester won the coveted Malcolm Baldrige National Quality Award, which it shared that year with Cadillac and Federal Express. (No other IBM division has won this award.) In looking carefully at the AS/400 saga, the traditional model of corporate "R&D" seems less pertinent than the locally situated efforts in "D&M," or development and manufacturing.

IBM's success with the System/360 locked up the mainframe computer market but created vulnerable spots—both above and below. In the upper reaches of supercomputing, IBM's mandate that its prototype ASC supercomputer be compatible with the 360's operating system meant that machine was never to touch the superfast Control Data and Cray machines; indeed, once this became clear, the ASC-360 was summarily canceled in 1969. At the lower end, IBM also recognized that it needed to make an entry in the burgeoning minicomputer market, where DEC and Data General were carving out impressive profits in selling smaller, cheaper computers. Even the smallest System/360 model 20—developed jointly at IBM facilities at San Jose, Endicott, and Rochester—was still far too large and expensive to compete with DEC's PDP-8 introduced in 1965 at eighteen thousand dollars.

In 1969, IBM launched the Rochester-developed IBM System/3. Rochester developed the new system entirely on its own and basked in the publicity of "the most significant IBM product announcement since the IBM System/360." System/3 used integrated circuits, rather than the transistor-based modules of the System/360, and for a thousand dollars a month it could be rented for half the cost of the smallest of the System/360. It was also a decisive break from the tabulating machine world. "From the very beginning a System 3 programmer never saw a card punch . . . a little bit rare I think," noted one its programmers. Over time, despite not being compatible with the range of System/360s, it sold more than twenty-five thousand copies, generating significant revenue and bringing significant internal and external attention to the "mid-range" market.[24]

Rochester continued its efforts to develop new "mid-range" computers as successors to the System/3, and introduced a succession of four new models between 1975 and 1984. The names were a confusing jumble, orbiting around the flagship System/360: System/32, 34, 38, and 36. It was perhaps indicative that System/36, which seemed to dangerously echo the flagship 360, actually came after System/38, which itself had been delayed. IBM Rochester sold large numbers of System/36, but at such large discounts that it never made money, while System/38 sold profitably but at inadequate volumes. Worse, in a way, the growing power of the upper range of these machines began to overlap with the lower end of the System/360 and its successor 370, and there were three additional middle-range IBM computers vying for the same space. "Here sit these five warring product lines, and the corporation now was saying this was a difficulty," stated one Rochester manager.[25] The five lines did not share applications, did not interconnect, and barely even communicated. For a rationally run corporation, it was something of a marketing nightmare.

In an effort to better organize its middle-range product line, IBM initially convened a project known as Fort Knox. It was a distributed, multi-site effort (1982–85) to design a single mid-range computer that could replace the confusion and jumble of multiple options with a single clear choice. In an attempt to promise "all things to all people," Fort Knox was a failed effort in corporate planning.[26] Its successor, Silverlake, aimed at creating a new product in the unheard-of time frame of two years. Partly, it was an act of desperation: five thousand people worked at Rochester, roughly half each in manufacturing and development, and with the System/38 that IBM Rochester had pinned its future on ominously declared "nonstrategic," the facility needed a new mainstay product—and soon.[27]

Silverlake was hatched in 1985 as a classic "skunk works" project, with internal secrecy and utmost speed being the paramount concerns. It is by no means clear whether IBM's top management knew of, let alone approved, this Rochester-based initiative. The head of IBM Rochester's development lab, Dave Schleicher, authorized a five-person team led by programmer-engineer Pete Hansen. Hansen assembled a hardware designer, an experienced product planner, an expert in advanced technologies, and a sales whiz. Remarkably enough, by the end of that year Hansen's team had outlined a plan for a successor mid-range computer and proposed a development schedule of two years; and, even more remarkably, the executives at IBM in due time eventually authorized the project's price tag at $1 billion. Tom Furey, the new development manager at IBM Rochester, helped sell the project to skeptics with an inspiring vision of world leadership in middle-range computers and hands-on attention to local concerns. There were still numerous technical and political hurdles. Hansen's group had effectively made a bet on System/38's technology, despite the rival System/36's having fully ten times the number of installations around the world. It was still unclear how to design and build six different computers, with projected prices ranging from $15,000 up to $1 million. Nor was it certain, especially with the famous software

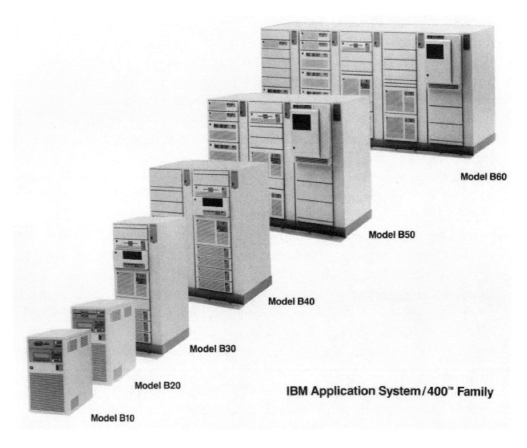

FIGURE 6.6. IBM Rochester designed and developed the Application System/400 family of midsize computers. Sales of these units brought hefty revenues to IBM's bottom line, including $14 billion in 1990 alone. Source: IBM.

disaster of OS/360 in mind, that a new operating system with seven million lines of code could be created in short order.[28]

Several organizational changes imposed by Tom Furey broadened the intellectual and political scope of the development effort. These changes facilitated interaction between development and manufacturing staff, creating horizontal communication in the largely hierarchical corporate structure. As development chief, he had no formal authority over Rochester's manufacturing effort, so he instituted a series of "management board" meetings where for a full morning each week the development and manufacturing staff could frankly discuss problems and iron out difficulties. He also sent a representative to corporate headquarters in White Plains, New York, as much to keep alert to upper-level corporate politics as to keep upper management apprised of Rochester's promising project. More controversially, he instituted "cross-functional" management within the Rochester workforce that, despite the skeptics, seemed to help constructively deal with challenging problems identified by current customers that had previously too often gone unresolved or even shunted aside.

More challenges were soon to come when prospective users entered the picture. In marketing the new middle-range computer, IBM set aside some of its well-practiced routines, such as listening most attentively to the largest customers such as Boeing or the biggest government agencies, while correspondingly discounting the concerns of the smaller customers. A new market vocabulary of "segmentation," "targeting," and "positioning" helped make sense of the cacophony of hypothetical customers with an unimaginably diverse set of needs. Exploring the health-care sector in detail, especially hospital departments, pharmacies, clinics, physicians' offices, and medical laboratories, the Rochester marketing team made solid estimates of likely sales figures as well as identified the reasons why prospective users actually bought computers. Overwhelmingly, they found, it was useful software and helpful services. A new conscious focus on consumer-driven innovation was in the making.[29]

In line with the marketing findings, software was a prime concern as AS/400 took shape. Rochester's programmers had already done such landmark packages as RPG II, a follow-on to the groundbreaking Report Program Generator program that had sold thousands of IBM 1401 computers to users migrating from punched-card equipment. Rochester's growing squad of programmers designed RPG II for the System/3 and its successors and also RPG III for the AS/400. While earlier versions still followed the "punch-card paradigm" of tabulating cards and assembling reports, RPG III was in effect a major programming language, such as BASIC or FORTRAN, complete with conditional testing, program loops, and subroutines. The need for applications software for the latest generation System/38 and System/36 machines also bore on the software designers. From a software perspective, AS/400 combined 36's large portfolio of applications and wide range of connectivity with 38's support for programmer productivity and integrated database. System/38 had "a database that was unique to Rochester. It was one of those deals that we can do it better than anybody else . . . Rochester has a lot of pride in what they can do," according to two participants.[30] Languages and utilities came from an IBM software development group in Toronto. A "layered" architecture helped manage complexity, effectively tying together the lowest layers of hardware with the upper layers of user interfaces. Some of the features, such as built-in support for relational databases, gained in significance and importance as the years passed, while other designed-in features, such as "tight integration" with IBM's short-lived OS/2 personal computer, did not.[31]

To realize these lofty ambitions, Rochester adapted two innovations that emerged as trendy management school topics: concurrent engineering and customer-driven development. They were related.[32] With the acute time constraints promised by Silverlake, AS/400 did not have the luxury of IBM's usual orderly sequential product cycle of planning, development, manufacturing, sales and service, with customer contacts coming safely at the end after all the important decisions had been made. Nor was the usual in-house secrecy, where customers were kept in the dark about new products until formally unveiled, any real advantage. One degree of concurrency occurred when the hardware engineers thought they might be able to cut the usual "two-pass" development cycle neatly in half, gaining a working prototype with just a single "pass" of design and development and thereby saving ten months. Extensive computer modeling of the hardware design—about the closest AS/400 ever came to traditional R&D—was the key to achieving an unusually reliable prototype, essentially straight off the drawing board. That prototype was passed directly to manufacturing, which turned out forty copies so that software development could start quickly. Manufacturing gained early experience that quickened its own transition to volume production. Meantime, the hardware engineering group that would have built the prototype instead moved on to designing peripherals and interfaces.

Soon enough, even as the hardware designs were stabilizing and while software was being written, AS/400 development was opened up to prospective users. Several hundred targeted customers, application programmers, systems engineers, and value-added resellers then responsible for 60 percent of IBM system sales were invited to attend "customer councils" and "migration invitationals" at Rochester (and later in Europe and Asia as well). The extensive external participation, including IBM's own field engineers, added external personnel worth an estimated one hundred person-years to the Rochester development effort. Especially given IBM's traditional secrecy with code names and secret documents, "opening up the doors of the laboratory and bringing in hundreds of application developers and customers, individuals, was really considered revolutionary." IBM's unusual attention led one customer-participant to exclaim, "Do you realize that 90 percent of what I suggested is now in the product? I feel like I'm part of the development team."[33]

Something as big as AS/400 could never be completed by a single IBM facility, no matter how talented. Months of preparation and planning preceded the public launch on June 21, 1988. In place were legions of sales representatives, poised to reel in orders, as well as some special precautions. A twenty-person "red team" including some of IBM's most experienced salespeople did their own in-house scrutiny of AS/400, and the claims IBM would be making for it, looking for weak points that rival companies might unduly magnify in the trade press. Even before the launch, there were one thousand applications programs already tested out and, according to one estimate, nearly five thousand early shipments of AS/400s already in customers' hands. A select group of technology journalists and industry analysts got a two-day advanced briefing that turned Rochester-based Bruce Jawer into a celebrity for his command of technical detail and calm manner. IBMers wore "Ask Bruce" buttons. Within four months of the splashy launch, sales had topped twenty-five thousand, exceeding even the brisk pace of early sales of the IBM personal computer after its launch in 1981. Rochester's full-court press to fix programming bugs and other problems attracted unusual attention. As one account put it, "I have a problem with the AS/400. I hate to report it to Rochester because as soon as I report a problem, within twelve hours, I've got three [Rochester] people in my lab area that want to work on this problem. It interferes with my ability to do work." Besides the financial rewards, IBM's Rochester division was twice nominated for the Malcolm Baldrige National Quality Award, winning in 1990. "We celebrated the news . . . in a prototypically Midwestern manner," as three Rochester insiders described it. "We had an old-fashioned ice cream social in the cafeteria."[34]

Supercomputing

Rochester arrived at the apex of supercomputing in an unexpected way. Roughly speaking, the three developments that converged were Rochester's need for a new line of technical activity, IBM's announced "grand challenge" to build a new supercomputer, and Lawrence Livermore's unbounded appetite for supercomputing muscle. It was a classic combination of motive and means. You might even see apt parallels in Rochester's casting about for a new product for which to deploy its accumulated technical talent (and to employ its existing staff) and the specialized manufacturing auxiliaries casting about and finding new markets in medical devices (see chapter 8). For a facility that had started off assembling electromechanical sorters and collators, it was quite an achievement to meaningfully mention Rochester, Livermore, and IBM grand challenge in one sentence. And, because it depended on extensive interaction with IBM researchers from Yorktown Heights, the Blue Gene supercomputing project also significantly extended Rochester's D&M model.

The revenues flowing in from IBM's successful launch of AS/400 in 1988 did not prevent the corporation from hitting a financial wall in the early 1990s. IBM had been famous for increasing its annual revenues, nearly every year since the economic expansion of the 1950s, the very model of a profitable and mature "blue chip" company. But its growth began stagnating in the 1980s and by 1991 it tipped unmistakably into the red, racking up immense losses of $8.1 billion in 1993. IBM had grown very large on the sales of mainframe and mid-range computers, but it was largely missing out on the personal computer industry. Although it had defined the personal computer in 1981, IBM had little experience in selling cheap commodities directly to consumers and the stellar profits from personal computers flowed first to Compaq and Dell and ultimately to Intel and Microsoft. (In 2004, IBM exited the personal computer business.) Personal computers first ate into the market for minicomputers, felling the once-predominant DEC, which was acquired by Compaq in 1998 in a merger that created the number-two computer supplier behind IBM. The rise of the Internet and World Wide Web in time created a new market for networked personal computers and large banks of servers, but in the meantime the standard model of mainframes and mid-range computing looked antiquated.

A new corporate-wide "services" strategy implemented in 1993 by Louis Gerstner Jr.—following the lead of H. Ross Perot's Electronic Data Services and others—was a distinctly mixed blessing for Rochester. On the one hand, it restored IBM's corporate profitability. By 2001, IBM's Global Services Division, with thirty billion dollars in revenue, or roughly 40 percent of the whole, was its largest division. Services were soon embraced by such erstwhile hardware giants as Hewlett-Packard and Xerox as well. (Control Data, as chapter 4 recounts, was perhaps just a few years too early in embarking on its own information services strategy.) On the other hand, in the new services-centric IBM, Rochester's proven expertise in manufacture, development, and software was moved to the margins; it is no accident that Rochester's peak employment in the early 1990s subsequently slid downhill to a recent figure around five thousand. "When Gerstner took over everything changed," according to two longtime IBM Rochester managers.[35] The first step that brought the Blue Gene supercomputer to Rochester occurred in 2002 when it formed an "engineering and technology services" group and began scouting around for a likely project. With the new services strategy "we [were] looking for ways to use the surplus of engineering skill. . . . You want to keep the people and keep the talent. Keep the skill base," stated Rochester-based Blue Gene project leader Steve Lewis. The Blue Gene project, it turned out, was a well-timed opportunity to "to make money by hiring out the engineering skills."[36] Here was a strategy, oddly enough at one of America's largest firms, that would also be followed by dozens of smaller and medium-sized firms as Minnesota's high-tech industrial district reoriented itself from computing to medical devices (see chapter 8).

The conceptual origin of Blue Gene can be traced to IBM's Thomas J. Watson Research Center, where computer scientists in the late 1990s were examining the next generation of supercomputing. A scientific challenge in the offing was the modeling of protein formation. IBM researchers thought that a significant science milestone could be achieved by a supercomputer with the speed of a petaflop, a million megaflops or floating-point operations per second. Such a machine, running continuously for a year, might completely model the folding of a complex protein. Existing supercomputers were still in the teraflop range, a thousand times slower. Pushing forward the frontiers of science while advancing the state of the art in technology—and gaining public renown for both at once—seemed perfect for IBM. It was a "grand challenge."

Advances in supercomputing from Seymour Cray's day onward required an intricate blend of fast hardware and novel architecture. For more than three decades, "Moore's Law" created successive generations of semiconductor chips that were ever-faster and smaller, but IBM researchers knew that a time was coming soon when the buildup of excessive heat would blunt this trajectory. Making chips smaller means that there is less surface area to release heat: think of a hundred-watt incandescent lightbulb the size of your fingernail. Making chips run faster was even worse for heat. (Owing to the looming problem of "heat death," Intel itself in 2004 shelved faster versions of the Pentium 4 in favor of a new hardware trajectory using multicore processors.)[37] Even though upper-end microprocessors at the time were breaking into the gigahertz range, or 1,000 MHz, IBM researchers built a prototype of Blue Gene with relatively slow 440 MHz chips. One key was in the architecture that permitted superfast networking between computing chips, so that the processors would never sit idle. The other key was in cutting the energy consumed by each processor, with the slower speeds, so that you could simply scale up in a linear fashion, adding additional cooling equipment as you added additional racks of the modular supercomputer. The result was that you could—at least for certain kinds of computing problems—increase Blue Gene's speed by increasing the number of processor cores. If you were rich enough, there was no physical limit on how many racks could be assembled into a supercomputing system.[38] When IBM Research built a "half rack" design with 512 compute "nodes," the prototype machine filled the top of a table.

A third key partner in the Blue Gene effort was Lawrence Livermore National Laboratory. In 2002, Livermore signed a contract to purchase the first sixty-four Blue Gene racks, assisting with the heavy costs of development as well as defining a number of key characteristics. Henceforth, in recognition of Livermore's crucial early support, the entire computer system was called Blue Gene/L.[39] While the research team needed only to get the half-rack prototype to work, the development team needed to set the size and weight of the individual Blue Gene "racks" to fit through doors, down hallways, and into elevators. The definition of height, particularly, "came from the interaction with Lawrence Livermore." The packaging experts at Rochester weighed in on the dimensions and weights that were needed for international air

FIGURE 6.7. Blue Gene development team at IBM Rochester, circa 2004. This group was responsible for the design, development, and testing of Blue Gene and manufactured sixty-four "racks" for Lawrence Livermore National Laboratory. Source: IBM.

shipments. Blue Gene's internal architecture, both hardware and software, also would be designed to solve the supercomputing problems that Livermore was concerned with, including its traditional activities in nuclear weapons design and stewardship as well as newer areas such as molecular dynamics and medicine.[40]

Blue Gene emerged from an unusual long-term working partnership between IBM Research at Yorktown Heights and the development team at Rochester. Normally, Yorktown Heights researchers only demonstrated concepts and prototypes: when they finished something like the half-rack prototype, their job was complete. "The real people part of this project is how do you get the development people involved early enough, long enough and hard enough to ensure a good end zone when the first half of the project has no economic value to your decisions? And how do you keep the research people on long enough to make sure all of the architectural bugs are gone when they have already collected their kudos?" observed manager Steve Lewis. Most unusually, to assist with this blended research–development task, a delegation of IBM researchers from Yorktown Heights took up residence in Rochester. Extensive in-person meetings between research, development, and Livermore representatives put members of the Blue Gene team on airplanes to California, Texas, and New York. There was the inevitable flurry of e-mails and instant messages. "But you have to have these points where you come together and look each other in the eye and say this is where we are at. . . . Are we still in agreement on this?" emphasized project leader Curt Mathiowetz.[41] Again, even in this era of extensive telecommunications, in-person contact, conversation, and concord were necessary.

Ultimately, Rochester's task was to scale up the "half-rack" prototype by two full orders of magnitude—to be precise: to make a working supercomputing system 128 times larger than the Yorktown Heights prototype. The operational machine would have sixty-four full racks totaling more than sixty-five thousand processing cores, all networked together to form a superfast machine—that would not melt down. To begin, think about the challenges of putting three hundred large one-hundred-watt incandescent lightbulbs in the space of a small closet. Roughly speaking, *each* of Blue Gene's sixty-four racks consumed this amount of energy: thirty kilowatts of electric power. When complete, the Blue Gene installation would consume energy enough to power a small city. It turned out that Rochester's experience in engineering development and high-technology manufacturing, refined with the AS/400 project and its successors, was a good match for the task at hand. Curt Mathiowetz recalled, "from a project management perspective, I worked with a dozen different people directly. They each had teams . . . on the order of a hundred people." "At the beginning they're all hardware designers; at the end they're all programmers," remembers Lewis. One of the challenges was balancing the imperative of speed and the desire for careful development. "In terms of normal computer development, this thing was fabulously fast," Lewis noted. With the novel computer-on-a-chip architecture, Blue Gene development went through an accelerated "two-pass" process. "At the end of the first pass, everybody says with six more months of testing we can really be sure that this is going to work when the second pass comes back," Mathiowetz stated. "And the answer is with six more months of testing, this project just got canceled because its financial life span ran out."[42]

To provide real-time access for the development team, four "racks" of Blue Gene were set aside for development work alone.[43] At a unit cost that was two to three times the production racks, this was a significant investment. These four racks wired together might have created the eleventh-fastest supercomputer, pushing aside such traditional heavyweights as the Pittsburgh Supercomputing Center, National Center for Atmospheric Research (Boulder), and even Cray's top-of-the-line X1. Precious hours of test time were "scheduled literally around the clock" on these four racks. Results from one test often had consequences for other components of the Blue Gene system. There was close interaction between the chip designers and the circuit-board and packaging engineers, because the multitude of components had to fit into the standard spaces. Heat and hardware were also yoked together. Even though the aim

FIGURE 6.8. Hand assembly of one "rack" of Blue Gene/L at IBM Rochester, circa 2004. Supercomputers ever since Seymour Cray's were a mix of high-tech engineering and one-at-a-time craft assembly. Source: IBM.

of Blue Gene was to use dramatically less energy, allowing the cooling problem to scale up in a manageable linear fashion, the cooling engineers still needed to ensure that the flows of air, piped upward from the floor below and flowing out through the top of the computers, actually did the work of keeping all the components in a Blue Gene rack at a safe temperature. One hot spot, anywhere in a rack, might cause a processor to overheat and slow down or fail. Other engineering teams needed time for electromagnetic testing and other refinements before the units were ready for manufacturing. "So those four racks had to do everything. It was a big task to manage the time slicing on those, and keep everyone happy," Mathiowetz recalled. Stability runs requiring long blocks of time were slotted for "the wee hours of the morning."[44]

By the summer of 2004, when the testing was largely complete and manufacturing was preparing to swing into full-scale activity, two deadlines pressed mercilessly. Lawrence Livermore was coveting the top spot on the Top 500 list of the world's fastest supercomputers, which had been occupied for three long years by the NEC Earth Simulator in Japan. Each half year, the Top 500 list is compiled and announced,

FIGURE 6.9. Blue Gene/L at Lawrence Livermore National Laboratory was the world's fastest supercomputer for more than three years (from November 2004 until June 2008), bringing Lawrence Livermore, IBM, and Rochester significant bragging rights. Source: Lawrence Livermore National Laboratory.

and the computer that lands at the top gains a hefty dollop of bragging rights. A year earlier, in November 2003, a single Blue Gene development rack was ranked seventy-third on the Top 500 list, while in June 2004 an eight-rack machine temporarily assembled at Rochester was ranked fourth.[45] So, simply by doing the math, LLNL knew that its full sixty-four-rack Blue Gene/L would claim the top spot.[46] The only question was by how much—and when? IBM's corporate leadership was mindful of the Top 500 list, but even more they were watching the financial repercussions. If Blue Gene could be shipped, and formally accepted at LLNL, by the end of the year, IBM's corporate finances looked better—not to mention Rochester's future. "Once someone can smell the money, it becomes very, very real," Mathiowetz observed. That summer of 2004, with the deadlines looming, the development engineers asked for three months of extra testing time to ensure the machine's stability, and finalize the software, knowing that the extra time would force the shipment to slide forward into March 2005. They were given three days. That fall, striving to send out the first round of sixteen racks to Livermore, recalled Lewis, "that's when we were, I would say, maximally stretched."[47]

Sometime that fall, engineers in Rochester created a supercomputer with sixteen Blue Gene racks wired together on the production floor. This "machine" does not appear in the official record books, but it was possibly the world's fastest supercomputer. This quarter-scale Blue Gene already might match the speed of the NEC Earth Simulator, and there seemed no insurmountable barrier to achieving a full sixty-four racks with a wide margin of speed. No one had time to rest on their laurels, and these sixteen racks were shipped to California for permanent installation at Livermore. Another sixteen-rack machine was assembled and tested in Rochester, then shipped out for Livermore. With the production line in peak form, Rochester was able to assemble and test thirty-two racks of Blue Gene, a record-setter that does appear in the books—if you look carefully.[48] "The seventy teraflops ... was probably the most momentous" turning point, remembers Lewis. Oddly, the sixty-four racks of Blue Gene were never tested in Rochester. The full sixty-four-rack machine, destined for no small measure of fame and glory, was first assembled on the Livermore floor.

Even with the sixty-four racks safely shipped out the door, there was still the nerve-wracking process of "acceptance" by Lawrence Livermore. This was no mere formality, as Mathiowetz recalled, because "all the pain was yet to come in getting through the acceptance processes." Blue Gene was nothing like an ordinary computer development project, such as AS/400 where there were nearly five thousand installed systems already at the formal launch event. With Blue Gene, not many spare copies existed besides the sixty-four racks that were in Lawrence Livermore's hands. And acceptance was an acutely pressing matter, because if Lawrence Livermore for any reason rejected Blue Gene, it was doubtful whether any other customer would accept the woebegone cast-offs. And, once again, no small matter of corporate finances hung in the balance. All of the partial payments Livermore had made to IBM, for deliveries of the racks as well as developmental milestones along the way, were classed by the accountants as a temporary matter of cash on hand. Only with the acceptance letter could IBM move these figures to the all-important income statement with accepted expenses and revenues. With "the sixty-four-rack acceptance" finally in hand, Lewis remembered a thundering outpour of relief: "Blood pressures all over the world drop!"[49]

In the years since the triumphant launch of Blue Gene/L, IBM Rochester has continued an impressive string of technical innovations and yet the changing corporate structure remains a challenge for any sense of local identity.[50] Those who know will credit Rochester for a string of innovations in superfast gaming chips, the product of its locally organized Engineering and Technology Services group. These are found in Sony's PlayStation 3, Microsoft's Xbox 360, Nintendo's Wii, and the newly launched Wii U. As noted earlier, there were ninety Rochester-built Power 750 servers

FIGURE 6.10. Rochester-built Power 750 servers formed the "Watson" supercomputer on quiz show *Jeopardy!* in 2011. Source: IBM press kit, www.03.ibm.com/press/us/en/presskit/27297.wss (July 2012).

at the heart of IBM's Watson supercomputer when it stole the show on *Jeopardy!* in February 2011. "I, for one, welcome our new computer overlords," noted one of the defeated human champions.[51] In comparison, IBM's official profile on its Minnesota facility is somewhat muted, emphasizing "IBM Rochester: Harvest in the Heartland" and the tabulating machine and early computer-manufacturing days. Remarkably enough, it carries Rochester's story only up to 2003—the very last line is "IBM's Engineering & Technology Services organization maintains a group in Rochester"— well prior to the gaming chips that resulted from this group and no mention of the Watson–*Jeopardy!* connection.[52] Perhaps, as elsewhere, it is a job for Minnesotans to identify their local heros and celebrate their successes.

7
Industrial Dynamics

Minnesota Embraces the Information Economy

Industrial districts the world over are dynamic and unstable spaces, always in motion and never staying the same for long. This was the case in the classic districts in Manchester or Sheffield during the British industrial revolution, the urban industrial districts set up in Chicago and New York that thrived during the heyday of twentieth-century manufacturing, as well as the latter-day high-technology districts of Massachusetts's Route 128 and California's Silicon Valley. Although communities and residents might prefer it otherwise, the centers of yesterday's money-spinning manufacturing in Newark or Detroit or Cleveland are too frequently today's forlorn industrial slums. MIT researchers pointed to regional specialization and "the explosion of new products" to comprehend the booming Massachusetts minicomputer industry in 1983 but the once-proud Data General and DEC, twin pillars of this notable success, are long gone.[1] The relentless pace of change has been a difficult lesson to learn in Minnesota, which has over the decades gained and lost prominent industries in saw milling, grain milling, woolen making, meatpacking, steelmaking, and automobile manufacturing. Ford began making Model Ts in 1925 at the Twin Cities Assembly Plant in St. Paul and closed up operations with Ranger pickup trucks in 2011.[2]

Minnesota companies were at the forefront of the computer industry for two or three decades, but it was a commanding position that proved impossible to sustain. Looking back, it is rather remarkable that Minnesota computing thrived across three distinct generations of technology—vacuum tubes, transistors, and integrated circuits—as well as wrenching changes as computing matured from its founding in government-subsidized niches and entered into the mainstream. These changes required engineers, designers, and managers as well as production workers and maintenance people to master new skill sets several times in one working career. This chapter

190 INDUSTRIAL DYNAMICS

FIGURE 7.1. Repairing vacuum-tube computer in 1953. This computer was built by Burroughs, which later merged with Sperry to form the Unisys company in 1986.

examines the tools that historical actors used to appraise "industrial districts" on their watch and manage the results as Minnesota's high-tech economy grew from a base in computing and diversified into medical devices and other areas that are emerging today. Over nearly forty years, local business leaders, government officials, university administrators, and interested stakeholders in Minnesota's high-technology industries variously measured, assessed, and debated the importance of local skills, state policies, finance, and education. The focus shifts to the dynamic sectors in industry, commerce, finance, and education that were key beneficiaries of Minnesota computing as well as essential early markets that shaped the emergence of the state's computing industry. Chapter 8 rounds out the treatment of Minnesota's high-technology economy today. My analysis of Minnesota manufacturing from 1980 to the present reveals some surprising results. Even though Minnesota's core computer companies have faded from center stage, the networks of specialist auxiliary firms and ancillary industries that grew up around them were a base for the state's medical-device industry. And they are actively reshaping the state's high-technology industries today.

Local Skills

Industrial districts depend on an unusual "skill intensity" evident in these highly dynamic regions. Farmers might be generalist jacks-of-all-trades, but workers at all levels in industrial districts often develop special skills and knowledge that are impossible to easily replicate elsewhere. Indeed, this is one of the chief reasons why companies locate in these hot spots of innovation despite their higher costs for labor and space. In addition, so-called knowledge spillovers typified by the presence of highly skilled workers in one firm or industry, with substantial interfirm mobility, can promote "horizontal" innovations within the district while raising the bar for competing firms and industries beyond it. Highly specialized managers, engineers, designers, and skilled technicians are cornerstones in the establishment and persistence of high-technology industrial districts.[3]

In Minnesota, a combination of state-level activities, supported by data gathered by the federal government, helped pinpoint the growth of a highly skilled workforce beginning in the 1950s. For a state traditionally dependent upon agro-industry (such as General Mills and Pillsbury), it was a profound shift to high-tech manufacturing and knowledge-intensive services. Skills built up in earlier phases of Minnesota industry were essential building blocks for later phases. Specialized metalworking and machine-shop skills were a building block for the computer industry, which created networks of supplier companies that extended Minnesota's sophisticated expertise in precision manufacturing. The firm-level evidence discussed in chapter 8 indicates that the region's prominent medical-device industry—now famous as Minnesota's Medical Alley—drew on these computer-sector supplier firms.

Many local communities and states across the country faced the booming postwar economy by seeking to apply lessons learned from the success of economic mobilization during the second world war. Striking the right balance between growth and change was not an easy task. As Roseville experienced already by 1960 (see chapter 4), there was the "growth dilemma" triggered by the great engine of suburban development, with its pronounced and expensive impacts on a community's requirements for streets, schools, water and sewer systems, as well as increased car and truck traffic generated by the newly created suburban shopping malls. The pioneering Southdale shopping mall, the first enclosed shopping mall in the world when it opened in 1956, competed for metropolitan shoppers' favor with the open-air Knollwood Mall in St. Louis Park (1955) as well as the enclosed malls at Apache Plaza in Minneapolis–St. Anthony (1961), Brookdale in Brooklyn Center (1962), and Rosedale in Roseville

FIGURE 7.2. Manual wiring of computer assembly at Control Data in the 1960s.

(1969). To anticipate and manage these dynamic structural and spatial changes, city and regional planners relied on detailed employment surveys and forecasts of occupational needs.

If you look carefully, you can see already in the 1950s the emergence of Minnesota's high-technology industry in the occupational outlooks and regional wage surveys. These surveys also make clear that not every Minnesotan shared equally in the benefits of these new highly skilled and highly paid job opportunities. In 1954 there was no separate entry for "computing" in the comprehensive *Occupational Wage Survey* that examined nearly a quarter million jobs in the metropolitan's industrial, transportation, communication, wholesale, retail, finance, and services sectors, but nonetheless there were a sizable number of early computing jobs.[4] Occupations that in a few years were transformed into full-fledged "computing jobs" included the operators of tabulating machines, billing machines, bookkeeping machines, Comptometer

FIGURE 7.3. Woman inspects microcircuit that enclosed complex wiring in 1978.

adding machines, and key-punch machines, as well as a growing number of computer designers and assembly workers in the manufacturing sector. Minnesota men figured most prominently in the tabulating machine sector, where—if wage data is an accurate indicator—there were two distinct levels. At the bottom was an ill-paid cluster with weekly earnings between $45 and $55, scarcely above the middle range of entry-level office boys. At the upper end was a second sizable cluster earning between $75 and $95 a week. These were the earliest "computing boys," who served in managerial or programming capacities. These high-skill workers were responsible for the intricate "setup" of the tabulating machines to do statistical analysis, financial computations, and other information-processing tasks. Once the tabulating machines were set up, the actual running of thousands of punched cards could be left to lower-skilled workers who monitored the routine operations. Inside a decade, many of these upper-tier workers would become the first computer "programmers" (a new occupation emerging in these very years). The region's specialized white-collar information-processing workforce offered a means, then, for Minnesota men to directly enter the computer industry.[5]

Minnesota women in 1954 experienced distinctive gender-specific employment. To begin with, the women operators of tabulating machines had access only to the lower-tier jobs. Even though many women (134 women and 268 men) worked as tabulating-machine operators, their standard weekly salaries were significantly lower ($56.50 versus $66.50 for men) and they were effectively closed off from the higher reaches of this occupation, where there would soon be significant (male) advancement into the ranks of computer programmers. Strikingly enough, just three women were in the upper tier of weekly earnings (over $75) compared with no fewer than eighty-one men. Minnesota women also experienced significant wage discrimination in the other proto-computing occupations: their standard weekly wages were below $50 in billing machines, standard bookkeeping machines, and key-punch machines, while women's weekly wages were slightly higher in Comptometer adding machines ($54) and the elite class "A" bookkeeping machines ($61.50). Again, the $50 weekly wage was just above the middle range of office girls, an unskilled entry position. Additional higher-wage occupations dominated by men included accounting clerks and order clerks, where comparable women's wages were lower by 20 percent or more.

So important were these insights into the changing labor market that the state of Minnesota stepped in when the Chicago branch office of the U.S. Bureau of Labor Statistics failed to keep pace. The state's *Occupational Outlook* published in 1966 defined the key economic region as the five-county metropolitan Minneapolis–St. Paul area, consisting of Anoka, Dakota, Hennepin, Ramsey, and Washington counties. By this time, the Twin Cities metro had become the undisputed center of a sprawling regional economy stretching from Montana across the Dakotas to northern Wisconsin and upper Michigan (indeed, the area of the Federal Reserve Bank of Minneapolis discussed below). The emergence of a major metropolitan center—with nearly 1.5 million people it was the nation's fourteenth largest—reflected some unsettling shifts in the region's economy. The Twin Cities emerged as the upper Midwest's regional center for retailing, wholesaling, meatpacking, and specialized manufacturing, in addition to its long-standing prominence in agricultural processing. Also during these years, however, the relentless mechanization of agriculture meant that "the youth

FIGURE 7.4. Women (possibly at First National Bank of Miami) operate Burroughs F-600 banking machines in 1960 as part of a gendered information-economy workforce.

of the region are finding it necessary" to abandon their families' farms and seek work elsewhere. Many of them gravitated to the Twin Cities. Whereas in 1940 the Twin Cities metropolitan region had 35 percent of the state's population, by 1960 it was already 43 percent (and today the comparable figure stands at around 60 percent). The in-migration of displaced farm youth also made the Twin Cities much "younger" by 1960—switching from being significantly older than the national average to being significantly younger than the national average. Displaced farm women became the feminized clerical workforce that underpinned the metro region's multitudinous information-processing activities. One hesitates to put it this way, but the region's decaying farm economy and the metro's rising high-technology economy were joined at the hip.

The 1966 *Occupational Outlook* clearly registered the emerging electronics and computer industry, lending statistical precision to the pioneering media profiles earlier in that decade (described in this book's introduction). Manufacturing, employing more than a hundred thousand people, was the largest of the industrial groups and represented fully one-fourth of the metro region's employment. The electronics and computer industry was a "new and important addition" to the regional economy, having "become one of the largest electronics manufacturing centers in the nation."[6] Employment prospects were bright for engineers, scientists, and professional specialists in manufacturing, and there was also "phenomenal growth in the employment of technicians" who worked side by side with graduate engineers and scientists. The five-year forecast was especially promising for electrical draftsmen (up 30 percent), electronics technicians (up 32 percent), mathematicians (up 42 percent), computer programmers (up 44 percent), and electrical engineers (up 50 percent), owing to strong expansion in these industries as well as the need for replacement workers. This was an auspicious time to be a computer "programmer," a new occupational category, whose work it was to "develop and prepare diagrammatic plans for the solution of problems by means of electronic data processing"—indicating the occupation's aspirations to be in charge of high-status brainwork. (There was, oddly enough, simply no mention of the essential labor-intensive work of writing computer code that enacted the programmers' "diagrammatic plans," let alone the time-intensive debugging and maintaining of computer programs.) These plan-oriented programmers worked in all sectors of the economy, with notable clusters in manufacturing (60 percent) as well as finance, insurance, real estate, and wholesale trade.

The metro area's lower-status clerical workers, where women held three-quarters or more of the jobs, were experiencing strong growth, too. By this time in the mid-1960s it was painfully clear that lower-status information processing was "women's work." Minnesota women accounted for 87 percent of billing-machine operators, 92 percent of calculating-machine operators, 93 percent of adding-machine operators, 95 percent of bookkeeping-machine operators, 99 percent of keypunch operators, and fully 100 percent of magnetic-ink encoder operators and bank-proof machine operators. Significantly, two clerical occupations soundly dominated by men were tabulating-machine operators (72 percent men) and computer console operators (80 percent men), where the five-year forecast of job growth, respectively, was 14 percent and a whopping 99 percent. In the vast sargasso sea of feminized clerical employment, the only other male-dominated clerical occupations were claims adjusters, insurance underwriters, and hotel room clerks.

The metropolitan area's blue-collar occupations were rigidly segmented by gender. Minnesota men were 100 percent of automobile mechanics and milling-machine operators (and typically dominated dozens of additional occupations), while women were 98 percent or more of nurse's aides, sewing-machine operators, and stewardesses. Some of the male-dominated blue-collar occupations were clearly related to

FIGURE 7.5. Teletype terminals, 1957. These terminals were used by Minnesota schoolchildren to run BASIC programs in MECC's statewide time-sharing computer system.

the electronics and computer industries (such as electricians, electronics repairmen, and machine-tool operators), but none of these blue-collar occupations can be wholly connected with electronics or computing. In the wider information economy, it was explicit that technological change was undercutting the traditional work of linotype operators, photoengravers, and stereotypers, resulting in stagnant employment where "most [job] opportunities will occur to replace workers who permanently leave the occupation."[7] It was still the heyday of directory, calendar, and newspaper publishing, though, and jobs for press feeders and pressmen had significant expansion (up 40 percent and 16 percent, respectively).

By the mid-1970s, Minnesota had two dozen computer companies that formed a major economic sector. Nearly all of the companies were located within the five-county metropolitan area (see table 7.1) with the only sizable exception being IBM's factory complex in downstate Rochester. In addition, General Fabrication had small assembly sites in Pine City and Forest Lake, and Control Data had one downstate assembly plant in Redwood Falls. This geographic concentration clearly indicates that the computer industry was one of the forces that concentrated the state's economy and employment in the Twin Cities metro. Although the computing industry was tabulated under the Standard Industrial Classification (SIC) for the business machine industry (SIC 357), computers by the 1970s were roughly ten times larger than the other subsectors of typewriters, accounting equipment, and office machines.

Minnesota's manufacturing, led by strong growth in computing and business machines, gave the state economy a massive shot in the arm. From 1960 to the mid-1970s, Minnesota's share of nationwide manufacturing employment went up by an astounding 50 percent (to 3.3 percent), with similar increases in manufacturing value added. At the time, Minnesota's computer manufacturers (SIC 3573) created nearly thirty thousand jobs and constituted a steadily rising share of the state's manufacturing workforce (8.7 percent). No less than nine firms employed five hundred or more workers, remarkable in that so many computing firms had been recently created as small start-up concerns and grown from there. Even more impressive was the state's share of the $6 billion national market for office and computing machines at around 12 percent, despite some cutbacks in military and NASA contracts that hit Minnesota hard. Likewise, Minnesota's computer-manufacturing workforce constituted fully 17.4 percent of the nation's total.[8] This was striking evidence of Minnesota's coming of age as a digital state.

State Policies: Regional Planning

At its origin, Minnesota computing depended on the omnipresent government support of computing. As we have seen, the intelligence and military agencies were persistent patrons of Minnesota computing from the earliest ERA-vintage Atlas computer, through the Cray-era supercomputers bought by the national laboratories

TABLE 7.1 Minnesota computer manufacturers, 1972

Company	Locations	Employment	Equipment Manufactured
Astrocom Corporation	St. Paul	9–24	Data sets (modems)
Banner Engineering	Hopkins	1–8	Data equipment, controls, printers
Comstar Corporation	Edina	25–49	Industrial process controls
Comten, Inc.	Roseville	100–249	Communication computers
Control Data	Minneapolis, St. Paul, Normandale, Redwood Falls	11,386*	Computer systems and services, card readers, optical readers/printers, computer terminals, peripherals
Data 100 Corporation	Minneapolis	100–148	Computer terminals (2 plants)
E M R Computer	Bloomington	250–499	Digital computers, peripherals
Electro General	Hopkins	9–24	Industrial electronic counters
Fabritek	Minneapolis	500–749	Memory core systems
General Fabrication	Forest Lake, Pine City	300–573	Computer subassembly, circuit boards
I C A	Minneapolis	100–249	Computer appliances
IBM Corporation	Rochester	2,000+	Electronic data-processing equipment
Jet O Matic Engineering	Minneapolis	25–49	Computer components
M D S Atron, Inc.	St. Paul	100–249	Computers, terminals, controllers
Memorex Corporation	New Hope	250–499	Computer system components
Metro Machine and Engineering	Bloomington	50–74	Machine tools, packaging machinery
Minnetonka Engineering	Hopkins	1–8	Computer machine parts
Module Electronics	Minneapolis	50–99	Electronic subcontracting
National Computer Systems	Minneapolis	100–249	Optical mark readers, commercial printing
Research, Inc.	Eden Prairie	100–249	Analog programming, digital controls
Rochester Datatronics	Minneapolis [sic]	9–24	Optical test scoring computer
Schott, Oscar and Company	Minneapolis	75–99	Power supplies, custom assembly
Univac (5 plants)	St. Paul, Eagan	7,850**	Electronic data processing, defense
Vector Engineering	Bloomington	25–49	Digital readout measuring systems

SOURCE: *Minnesota Directory of Manufacturers, 1972–73* (St. Paul: Minnesota Department of Economic Development, 1972). Data from SIC 3571 Electronic Computing Equipment. Employment figures given as ranges.

 * Control Data Minneapolis, 8100 Thirty-fourth Avenue (2000+), St. Paul Lexington Avenue (—), St. Paul Fairview Avenue (—), St. Paul University Avenue (75–99), Minneapolis–Normandale (—), Minneapolis–Meadowbrook Road (75–99), Redwood Falls (—), Minneapolis–Berkshire Lane (—). Control Data total from Frank Dawe, May 6, 1990, in CBI (80), series 9, box 30, folder 18: WCN day files.
** Univac St. Paul, West Seventh Street (2000+), St. Paul Minnehaha Avenue (500–749), St. Paul Prior Avenue (500–749), Eagan (1000–1999), Roseville–St. Paul Highcrest Drive (2000+). Univac total from Table 3.1.

and Univac's Naval Tactical Data System and the related FAA air-traffic-control systems, and even the IBM Rochester-designed and -built Blue Gene supercomputer built for the Lawrence Livermore National Laboratory. Minnesota computing was by no means atypical in its early reliance on military markets. It is not frequently mentioned in Silicon Valley today, but Santa Clara County "had the important advantage at that time of proximity to the emerging aircraft and space industry which was to become the major customer for semiconductors and, indeed, the only customer for integrated circuits for many years."[9] Even today, the U.S. federal government continues to be a significant player in the market for military-oriented systems integration. Many of the promotional activities, and a great deal of the market stimulus for computing can be traced to specific local instances of state activity, as this section outlines, including the Federal Reserve Bank of Minneapolis and the University of Minnesota, and then (in the following sections) several innovative networks of computer users organized into the Minnesota Educational Computing Consortium, as well as the varied efforts, over two decades, to bring a statewide Internet network to Minnesota. These demand-side actors were no less important than the computer manufacturers themselves in making Minnesota a digital state.

The upper Midwest region profiled in the state's 1966 *Occupational Outlook* was, of course, precisely the region of the Federal Reserve Bank of Minneapolis. It is worth highlighting that the Federal Reserve system itself created a coherent economic region stretching 1,800 miles across Montana, the Dakotas, Minnesota, northern Wisconsin, and upper Michigan. This economic region emerged as it did largely because the Federal Reserve system collected reams of statistics about it (analyzed and distributed on punched cards and computer tapes) and fostered a sense of regional economic identity through hosting meetings, conferences, and workshops as well as forecasting economic activities on a regional basis. Montana bankers, and their local customers, were meaningfully connected to the Twin Cities, not least through the singular Helena branch office of the Minneapolis Federal Reserve. When the national Federal Reserve system was created in 1913 one might easily have imagined Montana and the Dakotas better fitting with the nearby Kansas City–based region, while railroads at the time clearly tied Minnesota and Wisconsin to the Chicago-based region. (Chicago's commercial dominance is no fantasy: for twenty years, Minnesota's commercial Internet was provided through connections in Chicago, as described later in this chapter.) The Federal Reserve's twelve districts were originally apportioned by population, although today the Minneapolis Fed has the smallest population of any of them. The upper Midwest regional structure was consequential in a number of ways.

The Federal Reserve Bank of Minneapolis itself became a significant force in the information economy. Its wide-ranging data collection, analysis, and dissemination activities increased the number of tabulating-machine operators needed in downtown Minneapolis in the 1950s and the number of computer analysts, programmers, managers, and maintenance workers needed ever since. And because the Federal Reserve system played an incalculable role in bringing a uniform information ecology to the nation's far-flung banking system, Minneapolis became a prominent regional node. As historian James Cortada emphasizes, the Federal Reserve pioneered the electronic transfer of money even in the prewar telegraph-based era and it set up the pioneering Bank Wire system in 1950 and subsequently embraced paperless electronic means for moving money at all levels of the banking industry, from personal checks and credit cards up to the largest transfers between banking systems. Computer-based telecommunication systems soon emerged to permit interbank transfers on a regional basis, and by 1970 the Federal Reserve opened its computer-automated central clearing facility in Virginia. A complementary international system emerged to deal with electronic transfers with banks beyond the United States. The Federal Reserve, according to Cortada, played a key leadership role in the "actual implementation of [computer-based] networks."[10]

The Federal Reserve Bank of Minneapolis also developed unusually close relations with the University of Minnesota's data-driven economics department. Among many other prominent faculty, Walter Heller was a key economic adviser during the Kennedy and Johnson administrations and he subsequently built the department into one of national prominence. Economics, of course, was one of the earliest of the social sciences to enthusiastically adopt mathematics, statistics, data processing, and computing. Heller connected economics professors at the university with researchers at the Minneapolis Fed in the 1970s. He hired up-and-coming stars from the best universities around the country, enticing them with the promise of direct interaction with the data-rich Minneapolis Fed. By far the most important result of the university–Fed collaboration was its significant contributions to the currently dominant "rational expectations" school of data-driven macroeconomics. Among this notable group were Neil Wallace, Thomas J. Sargent, Christopher A. Sims, and Edward C. Prescott. Prescott, who won a Nobel Prize for related work in 2004 (as did Sargent and Sims in 2011), taught at the university for nearly a quarter-century, and today continues work at the Federal Reserve Bank of Minneapolis.[11] Heller also hired Leonid Hurwicz, yet another economics Nobel laureate, who taught at the university for fifty-five years. His Nobel-cited specialty was in "mechanism design," or the devising of effective practical mechanisms to deal with imperfect markets. Ordinarily, one doesn't expect to find applications of Nobel Prizes directly in daily life, but Hurwicz, a lifelong politically active member of the Minnesota Democratic-Farmer-Labor Party, invented the "walking subcaucus" used by Minnesotans to nominate candidates. It is difficult to imagine this churn of economics—including four Nobel laureates—occurring in Minnesota if the upper Midwest Federal Reserve had been located anywhere else.

The Minneapolis Fed also developed, like its counterparts, into a massive user of computing networks to move money in all its varied forms. Just to give some sense of the scale of operations, more than fifty computer technicians were called in to restore the Minneapolis Fed's computer system after it was knocked out by a burst water pipe on April 8, 1991. At risk was the ten billion dollars that moved—each day—through the Minneapolis Fed's Automated Clearing House (ACH) system for wire transfers, Social Security payments, payrolls, and daily credits to the vast network of more than 750 commercial banks across the upper Midwest. Indeed, the Minneapolis Fed district "differs from other parts of the country in the large numbers and wide dispersion of small, independent financial institutions that share the financial services markets with the district's larger institutions." In the 1990s, the Federal Reserve consolidated its twelve regional computer centers into three national centers, presaging a shift from regional to national scales that affected computing as well as banking. Data processing, statistics gathering, macroeconomics theorizing, and electronic funds transfers—all these were computing-intensive activities that the Minneapolis Fed concentrated in the Twin Cities.[12] The regional information economy that directly stimulated the state's computing industry had broad reach, massive volume, and deep roots.

State Policies: Educational Computing

The massive computing complex that energized and enveloped Minnesota industry, commerce, and finance extended into the state's educational system. The Minnesota Educational Computing Consortium (MECC) was a pioneering initiative to bring computing resources to the state's entire educational system, ranging from the upper levels of the university right down to elementary school students practicing their times tables on computers. MECC passed through two distinct phases. In its early years, it succeeded in bringing computing to 96 percent of the state's schoolchildren, providing computer hardware, networks, software, services, and training. For years it ran the world's largest educational time-sharing computer system, and then made a quick switch to the first generation of microcomputers, providing massive early orders to two struggling California garage-shop entrepreneurs. These Minnesota invoices helped make Apple Computer, Inc. into the country's leader in educational computing. Then, in its second phase as a state-held for-profit corporation, MECC became a nationally prominent developer and provider of educational software and courseware. It developed hundreds of software packages. MECC's "Oregon Trail" computer game, originally launched in 1971 as a terminal-based mainframe application, sold more than sixty-five million copies in a remarkable forty-year run. Today, even if you don't happen to have a terminal or mainframe, you can still play "Oregon Trail" because it has been ported to run on Apple's iOS, BlackBerry handhelds, Windows Phone 7, and even (for a time) on the social-networking site Facebook.[13]

In several respects, the improbable success achieved by MECC suggests it was a unique, could-not-happen-anywhere-but-Minnesota story. The educational activists who launched MECC, as well as the state legislators who supported it, clearly understood that Minnesota was unusually fertile ground—owing to the region's booming computer industry. In the early 1960s, a small group of progressive educators connected with the University of Minnesota's laboratory high school (fondly known as "U-Hi") began scheming about how to bring educational computing to their students and to the entire state. The at-hand pervasiveness of computing was an immediate inspiration: "at the time Minneapolis–St. Paul was the computer capital of the world; this is where all the action was [because] we made most of the computers here," one of the activists recalled, specifically citing Univac, Control Data, 3M, and IBM Rochester. "In the Twin Cities it was impossible to be unfamiliar with what was going on in the world of computing."[14] What followed in the next several years, leading up to the founding of MECC in 1973, is a classic tale of opportunistic bootstrapping and homegrown enthusiasm intertwined with industry support from Control Data and Univac.

The Minnesota educational computing enthusiasts at the University High School quickly located significant sources of local backing. One of the earliest was a Control Data programmer named Robert "Doc" Smith, who liked their idea of giving high school students access to computing.[15] The students would mail him selections from a library he provided them of pre-punched computer cards, creating a program, along with their own handwritten data. Smith somehow finagled company resources to punch cards with the students' data and submit the entire batch to one of the company's computers, returning the results to the students by mail about a week later. Frustrated that few of the student programs ran correctly first time out, the high school educators looked about for more immediate access. They came very close to snaring an inexpensive Univac 422 computer, built locally in Roseville, aimed at the emerging market for "colleges, universities, and vocational schools [that] are developing regularly scheduled courses in data processing." Instead, they took the prescient advice of Bob Albrecht, another Control Data programmer-analyst, to investigate a revolutionary experiment then under way at Dartmouth College in New Hampshire. Albrecht was a notable advocate for what became known as personal computing—in a few years time he moved to California and founded the countercultural *People's Computer Company* newspaper and the legendary storefront People's Computer Center in Menlo Park—and he was another of the free spirits that Control Data abetted and encouraged. Control Data paid Albrecht to drum up enthusiasm for computing among high school students and teachers, and it was while he was assigned to the company's Minneapolis headquarters that he heard about the Dartmouth experiment.[16]

The computing model that took form at Dartmouth in 1964 was to all appearances designed for the Minnesota computer enthusiasts. It was created by John Kemeny, a Hungarian-born math professor who during the second world war had been Albert Einstein's research assistant at the Institute for Advanced Study at Princeton. Kemeny had been introduced to computing after the war at Los Alamos and RAND. At the time, he wanted to teach programming to Dartmouth's students, who were not especially of a technical turn of mind. He had a vision of "millions of people writing their own computer programs," and his experience with FORTRAN, IBM's stolid and severe programming language (it was named for *Formula Translation*), was not encouraging. Inspired by a suggestion of MIT's John McCarthy, who was a former Dartmouth faculty member—and, as it happened, a pioneer in time-sharing and artificial intelligence—that "You guys ought to do time-sharing," Kemeny hired student-programmers and somehow created a workable time-sharing operating system that ran on a donated General Electric computer. Each user, working at a teletype terminal, could have immediate and direct interactions with the computer—no punched cards, no batches, and no lengthy waits. As soon as word of this feat got back to General Electric, it flew the students out for a hush-hush debriefing with its professional programmers to capture the principles behind their homemade time-sharing system, which GE soon elaborated into the core of its own commercially successful line of time-sharing computers. "Kemeny's Kids" also devised a basic programming language, soon known as BASIC, with a friendly and forgiving appearance. On May 1, 1964, at four in the morning, Kemeny and an undergraduate student each ran a short BASIC program (possibly the simple statement "Print 2 + 2") simultaneously over the Dartmouth Time Sharing System. This model of computing—impressively interactive and astoundingly personal for the time—was accessible across Dartmouth's campus, wherever there was a terminal, and also beyond campus through the General Electric computer's external ports.[17]

Armed with Albrecht's well-placed advice, the Minnesota group boldly cold-called Kemeny. It turned out that he had extra computing power at hand with the large GE machine and he offered them free access, so long as the Minnesota users could connect to it over long-distance phone lines. They snagged a five-thousand-dollar grant from the GE Foundation to purchase a standard-issue model 33 teletype and an acoustic coupler that would translate the teletype keystrokes into audio signals that could be sent over the phone lines, with the balance of around three thousand dollars to cover the long-distance phone charges. The ungainly teletype machine was literally put into the classroom, becoming one of the first-ever computer-in-the-classroom experiments (other early efforts were in Philadelphia and Stanford). Minnesota high school students thus gained direct access through Dartmouth to one of the leading experiments in computer time-sharing in the world. The mode of distributed remote access provided a model that they would use, soon enough, to wire the entire state of Minnesota.

Although other early computer-education efforts followed a "drill and practice" model, the Minnesota experiment soon moved forward in creating innovative and interactive content. "We believe that if kids in mathematics are given the opportunity to design a program," the educators held, and engage in the process of problem solving, they would grasp the central math concepts and gain "a skill whose usefulness extends beyond pure mathematics into other sciences, word processing, games, and other areas of life."[18] The Minnesotans' five-year effort from 1963 to 1968 resulted in the multivolume *Computer Assisted Mathematics Programming* textbook series by the nationally prominent Atlanta-based educational publisher Scott, Foresman. After two years of long-distance computing, and phone bills mounting, they needed a replacement for the Dartmouth machine. "Guess who had a GE computer? Well, in Minneapolis Pillsbury did," and so the students continued writing BASIC programs for a time-sharing computer, once again thanks to the local computing resources.[19] While the Minneapolis city schools also bought slices of time from Pillsbury, the suburban schools in 1968 set up Total Information for Education Systems

Gadgets linked to computers found pioneering uses in stock-quotation boards, airline reservations systems, defense networks.

Simplified input-output terminals open new ways to use information in offices, police departments, motel operation.

PRODUCTION

Computer's little helpers multiply

Thanks to a rapidly expanding line of simple, low-cost appliances,
you don't have to be a technician to use a computer now.
With phone hookups, you don't even have to be in the same city

FIGURE 7.6. Strong public awareness of computing in Minnesota in the 1960s led to strong state funding for MECC.

FIGURE 7.7. Control Data Cyber system (here in 1978) became the central computer for statewide MECC.

(TIES), a spin-off of sorts from the University's College of Education, to provide instructional and administrative computing to sixteen suburban school districts.

It so happened that companies like Digital Equipment Corporation and Hewlett-Packard were bringing out minicomputers at a fraction of the cost of a GE mainframe. TIES, expanding to cover forty-four school districts, bought an HP 2000 that could connect sixteen users simultaneously from the hundred or so terminals distributed across the suburban schools. Minneapolis followed suit with an HP machine, and there were similar experiments in St. Paul. Statewide, there were a total of five HP 2000s, a Univac 1106, and a Control Data 6400 linked up through a patchwork of time-sharing systems.[20] An impressive campaign was taking form. Decision makers in the schools were well steeped in computing: "When we went to see a school board . . . Guess who would be sitting on the Board? One person from Univac, one from CDC, somebody from Honeywell and somebody from 3M." Based on this unusually well-informed and solid support, "Minnesota was leading the way toward what was to become a major movement in the use of computers in schools and in the home."[21]

In this unsettled time, when models of educational computing and much else were in flux, Minnesota's governor Harold LeVander did a careful analysis and then set the wheels in motion that would create MECC in 1973. LeVander, a Republican, was a politician during a time in Minnesota's history very different from today: he created state agencies for pollution control and human rights, expanded voting rights, and imposed the state's first sales tax. He also noticed that the widespread enthusiasm for computing went straight across the myriad state agencies, prisons, and schools that were buying terminals and purchasing time-sharing services or computers outright. The governor required each state division to create and justify a strategic plan for computing, and from the state department of education came a stunning vision of putting computing within reach of every school in the state. The unconventional agency chartered with this mission would be the Minnesota Educational Computing Consortium, or MECC. It was created under the state's "joint powers" law, which meant that MECC itself inherited any power that each of its constituent agencies possessed. Because the University of Minnesota, the separate Minnesota State University System, and the Minnesota Department of Education could all employ people and purchase computers, so too could MECC. For arcane reasons, MECC also included the state's department of administration. MECC's sixteen-member board gained the power of approving—or denying—all educational computing budgets, even (formally at least) that of the University of Minnesota. MECC used these powers and its $3.6-million (two-year) appropriation to create a unified statewide system shaped by the vision of widespread educational computing.[22]

MECC in 1975 announced an industry-wide bid for a 435-port time-sharing computer, enough computing power to connect simultaneously to every school district in the state. MECC essentially took the Dartmouth-style model and extended it to the state, where "a giant computer in the Twin City area . . . serviced the whole state, kindergarten through graduate school." IBM, Control Data, and Univac submitted bids for the system. Despite Control Data's already having a promising computer installation at the university, its bid for the time-sharing system was fifty-five thousand dollars more than Univac's offer of a model 1100 computer, one of the last large-scale computers to use magnetic-core memory. MECC, under state contracting law, was compelled to accept the lowest bid—on the multimillion-dollar contract the difference was around 1 percent—and so the giant Univac was installed in fall 1976 at the university's off-campus Lauderdale computer center near Broadway and Highway 280. It was a risky gambit for Univac to take its batch mainframe machine, with orders rolling in from Lockheed, the air force, Shell Oil, and the University of Wisconsin, and convert it to a full-blown time-sharing one. Univac kept a crack crew of twenty programmers on-site, but they could not achieve the contract's stringent specification. After throwing out the troubled Univac machine in June 1977, MECC installed twin Control Data Cyber 73 computers, which were an updated version of Seymour Cray's legendary CDC 6600 supercomputers. MECC's time-sharing system, or MTS, provided full-day access to computers for 96 percent of Minnesota's schoolchildren,

FIGURE 7.8. MECC courseware in 1986 signaled the educational venture was branching out from its Minnesota roots.

and 80 percent or more of its schools, and was nationally hailed as a pioneering model. The Control Data system was operational 99 percent of the time from seven in the morning until eleven at night. "Minnesota leads the nation in computing activities in its educational system," according to a 1982 report by the U.S. Office of Technology Assessment. MTS, with its 448 active user ports, four-second response time, and two thousand terminals system-wide, was "the world's largest general-purpose educational time sharing system."[23]

Having been quick off the mark, MECC expanded its content-creating efforts and soon had an extensive library of programs for drill-and-practice routines, simulations and early computer games, problem-solving exercises, and various tutorials that it gave away at cost to the Minnesota schools. It seemed to be a perfectly successful implementation of a top-down central-computer model for education—until one MECC instructor, Kent Kehrberg, went to California and saw one of the earliest Apple II computers. The problem was that the compact desktop unit did nearly everything that the elaborate Control Data time-sharing system could offer, and more. Astoundingly, it could do high-resolution *color* graphics. The Apple II was the joint product of Stephen Wosniak, the technical genius who coaxed miracles from a cheap microprocessor chip, and Steve Jobs, the promotional genius who would—over time— create an information-technology juggernaut from these humble beginnings. They were desperately looking for orders at the time, hadn't a clue about educational computing, and listened attentively when the Minnesotans came to call. In short order, they dispatched to MECC five hundred Apple IIs, then, within three years, two thousand Apple IIs, and when all was said and done, the stupendous number of ten thousand Apple IIs. Apple gave MECC an attractive package price of $2,320 that included 32 KB of memory, a mini floppy drive, and color television.[24] By 1980, "MECC became the largest seller of Apple computers . . . Apple got its start in the educational computing business through its Minnesota connection."[25]

The Minnesota–Apple connection had several ramifications for the national computing scene. It is difficult to recall in our era of iPhones and iPads that Apple really grew from a dominant position in educational computing, far more than the home or business or entertainment markets. MECC got the microcomputing bug at an auspicious moment. Apple IIs went on sale in June 1977, and the very next year MECC began bulk purchases of the new machine. Because total worldwide sales of the Apple II by 1980 were one hundred thousand, without question the large orders from Minnesota created an important early market. Apple also benefited mightily because MECC was ideally situated to dominate the educational market. It simply converted its extensive library of BASIC programs and courseware to run under Apple's BASIC. Wosniak, in yet another tour de force, had put his own IntegerBASIC on a ROM chip in the earliest Apple IIs, and soon enough Apple licensed a version of Microsoft BASIC also squeezed onto a ROM chip. Switch on an Apple II, and it ran BASIC programs. Schools keen on microcomputing could purchase Apple IIs and a complementary suite of educational content from MECC. In 1981, MECC also began smaller bulk purchases and similar software support for Atari.[26] While it continued to provide software at cost to Minnesota schools, it developed an influential "membership" model that provided accessibly priced software to schools around the world—by 1982, there were already more than fifty such members across the United States and around the world, including England, Australia, Kenya, Saudi Arabia, and Switzerland—and with these numbers MECC gained a serious source of extramural income.

MECC began to look and act more like a for-profit software developer and somewhat less like a state agency responsible for providing computing in the four corners of the state. In 1983 MECC was rechartered as a profit-making taxable entity wholly owned by the state of Minnesota. Its handsome revenues for selling software and courseware were plowed back into developing more software and courseware—because there were no profit-taking shareholders—and soon MECC completely

FIGURE 7.9. Control Data's full-time Minnesota employment, 1971–89. Source: Frank Dawe, May 6, 1990, in CBI (80), series 9, box 30, folder 18: WCN day files.

dominated the Apple-based educational market. In the mid-1980s it offered three hundred software titles and had annual revenues of seven million dollars, making it one of the four largest educational software developers at the time. Its flagship "Oregon Trail" computer game, originally developed by three Carleton College student teachers two years before MECC was born, became a runaway success.[27] You chose characters from the legendary wagon trains in the 1840s that went out to Oregon Territory, facing starvation, dysentery, dying oxen, and other ailments and challenges. Early versions of the game tested your ability to quickly type "BANG" or "POW" when taking a rifle shot at a deer; if you typed fast enough, your family got venison for dinner. Soon, with the Apple II's graphics, you could see notable landmarks along the way. Oregon Trail spawned an entire series of adventure-simulation games, including the Yukon Trail, the Amazon Trail, and Odell Lake (named for an Oregon mountain lake), which pitted you, playing one of six different species of fish, against the perils of predators and fishermen. Oregon Trail provided approximately one-third of MECC's annual revenues of $30 million in the mid-1990s, after it was spun off from the state and sold to a venture capital firm for $5.25 million. In 1995, MECC, then a public company, was first bought by a Boston-based software house for $370 million, merged with Brøderbund and Mindscape, and then folded into a $3.8-billion package deal with toy behemoth Mattel, Inc.

Crisis at the Center

The early 1990s was not an auspicious moment in Minnesota computing. The state's most notable computer comany, Control Data, a high-performing stock-market darling for more than two decades, hit the severe bumpy patch described in chapter 4. Its employment peaked in the early 1980s, then fell precipitously (see Figure 7.9). By 1990, CDC had sold off several promising components, including its peripheral products division, its long-standing investment in the PLATO system, and the network of Control Data Institutes. Its loss-making supercomputing subsidiary ETA Systems was shut down with a painful $490 million charge to the company's finances. And then, with the retirement in March 1990 of Robert Price, who had joined Control Data three decades earlier as a mathematician–programmer, there was a changing of the guard. Lawrence Perlman, a lawyer specializing in corporation securities law (with a five-year stint as general counsel and vice president at Medtronic) took over as CEO. With Perlman at the helm, Control Data was run, for the first time, by someone clearly outside of the technical realm of ERA–Univac. Preparations were under way to split the corporation into two entities: Control Data Systems was left with the remaining bits of the computer business while Ceridian Corporation, under Perlman, took up its services businesses. As Perlman characterized his achievement a decade later, seemingly without irony, "We no longer make anything you can drop on your foot."[28]

FIGURE 7.10. ERA briefly memorialized, 1986–91. The plaque reads: "Engineering Research Associates, the forerunner of Sperry's Minnesota presence, is the acknowledged parent of some 100 Twin Cities computer firms. In commemoration of the 40th anniversary of ERA's founding, this plaque is placed on the company's original manufacturing site this 19th day of August 1986." The plaque vanished after Sperry left the plant in May 1991. Source: VIP Club vipclubmn.org/facilities.aspx (October 2011).

It turned out that 1991 was the final year for the once-revered Univac plant on Minnehaha Avenue in St. Paul. It was a long story and, in the end, painful. The Sperry Rand Univac operations in Twin Cities—described as the company's "largest in the world"—in 1968 employed 10,500 in its principal factories and facilities in St. Paul, Eagan, and Roseville, and its Minnesota spending on taxes, purchases, and payroll totaled $120 million. Even though business school gurus in the 1960s made the "conglomerate" business strategy all the rage, by the 1970s many of the resulting mergers turned out to be unwieldy and unprofitable. Indeed, the forcible unwinding of these ill-conceived conglomerates formed an entire chapter of American business history in the 1980s pivoting on the junk-bond or leveraged-buyout specialists. (Gordon Gekko—a composite of the era's high-flying financial buccaneers—in the 1987 film *Wall Street* declares: "Greed, for lack of a better word, is good.") During the 1960s, Sperry Rand, once focused on high-technology gyroscopes and computers, had assembled a sprawling set of disconnected holdings in typewriters, farm equipment, electric razors, and marine communication, as well as computers and military electronics.

INDUSTRIAL DYNAMICS 213

In 1978, after divesting these oddball divisions and dropping "Rand," it reorganized as Sperry Univac. In 1986, Sperry Univac merged with the venerable Detroit-based Burroughs company to form Unisys, which underwent further reorganizations. Sperry Flight Systems and Sperry Defense Products were sold, respectively, to Honeywell and Martin Marietta (eventually, Sperry Defense became part of Lockheed Martin). Sperry Marine in Virginia, the principal site for antenna couple manufacturing since 1970, became a division of Northrop Grumman. In 1986, just before the Burroughs merger, Sperry celebrated the fortieth anniversary of ERA's founding by installing a handsome commemorative plaque at the St. Paul plant. Five years later, in 1991, the aging plant was permanently closed down and the lease of the navy property abandoned. In the years since, the commemorative plaque has vanished. Its whereabouts remains a mystery.

With these ominous events pressing down, Minnesota governor Arne Carlson convened a top-level task force on "the future of the Minnesota computer industry." A preliminary report, issued in January 1991, seems to be its sole product. You need

FIGURE 7.11. Mark McCahill at the University of Minnesota developed the Internet Gopher, 1991–95: "the web before the World Wide Web." A multimedia Internet-based file sharing system, it had more than nine thousand servers around the world when the World Wide Web took off in the mid-1990s.

to read the report carefully; what is hidden is just as important as what is present. The group's two dozen members included Earl Joseph (a 1951 math graduate of University of Minnesota, longtime computer design engineer at Sperry Univac, and notable "futurist"), representatives from IBM, Unisys, two venture capital firms, the university world, and several high-tech industry, labor, and professional groups, as well as politicians, judges, and consultants. It was chaired by Carl Ledbetter, who had recently been though the wringer as president and CEO of Control Data's ill-fated supercomputing subsidiary.[29] Governor Carlson framed the group's charge as assisting "existing computer technology firms to preserve and expand our vital computer industry base." The centerpiece of the task force's recommendations—in addition to the predictable suggestions for tax breaks, incentives for entrepreneurship, and efforts to strengthen education (all laudable)—was a statewide "grand challenge" that harked back to the glory days of MECC. Sadly, nowhere did the task force appreciate the multilevel industrial district described in this book. The task force enumerated fourteen technical research fields, such as supercomputing, parallel processing, artificial intelligence, and computer architectures, where Minnesota had "world-class expertise," but notably absent from this list were the state's (equally) world-class resources in sophisticated high-precision manufacturing, flexible engineering and design, and extensive supporting ancillary industries. All the same, these largely unsung smaller and medium-sized companies were paving the way for an impressive medical-device industry.

The task force was, as these things often are, dominated by the center. You don't see "Sperry Univac" on the membership roster, even though this was the background of Earl Joseph, Francis Stephens, and University of St. Thomas faculty member Bernice Folz. Folz worked as a software development manager at Sperry Univac (and before then as a systems engineer for IBM). "Control Data" was also absent, but in reality it was well represented by a former vice president (Jane Belau, then a consultant); a former divisional president and CEO (Ledbetter, also listed as a consultant); a former computer engineer (Max Goldberg, then at IEEE and the city of Minneapolis); and one of its external accountants (Timothy Flynn of KPMG). Other center firms at the table were IBM and 3M. Supercomputing, too, was well represented. Greg Smaby was an analyst of the Minnesota supercomputing industry, and a frequently quoted expert on Seymour Cray. Donald Austin had been a computer scientist at Lawrence Berkeley Laboratory and the Department of Energy, before coming to the University of Minnesota as director of its Army High Performance Computing Research Center.

A bold vision for a statewide "information superhighway" to be known as MINNLINK-2000 became the task force's core recommendation. It suggested the project as the state's "single focus" for integrating computing and communications. As a state-level challenge, it was coupled with calls for strengthening K–12 education, math standards, and teacher training. The University of Minnesota was charged with building computer science into a "top-ten" nationally ranked department by 1995 as well as establishing a center of excellence in applied information technologies. Soon, these came to pass: the Digital Technology Center was established in 1998 by President Mark Yudof, and the university's computer science and electrical engineering departments have achieved top national rankings.

Implicitly, the task force seemed to see only the established anchor firms when it recommended MINNLINK-2000 owing to the "tremendous opportunities for the Minnesota computer industry." The challenging technical areas it outlined, in creating a statewide broadband fiber-optic network, included distributed networks, databases, software applications, and telecommunication systems. Chairman Ledbetter subsequently carried this vision to AT&T Consumer Products, Sun Microsystems, and networking specialist Novell in top-level executive positions for these companies. Oddly enough, however, especially for 1991—when the Internet was eight years old and already a major force in computer networking—the word *Internet* never once appears in the task force's report. Instead, it pointed to seven other state-level networking models in California, North Carolina, and elsewhere that it considered promising, as well as international models in Japan and France. "The French Minitel system, for example, links every home in France by means of a terminal attached through the phone system."[30]

The governor's task force in 1991 also missed an opportunity to connect with the University of Minnesota's Internet Gopher. It was world-renowned at the time—literally, it was the Web before the World Wide Web took its present shape. Gopher was the brainchild of Mark McCahill, a talented chemistry-and-computing expert. He had worked on the microcomputing side of the university (much in the shadow of the "big computer" people), programming Apple IIs and Control Data's Cyber computers. He and his team of programmers created one of the first client applications for accessing e-mail (POPmail is widely used even today by such providers as Earthlink). Released in 1991, Gopher was the first Internet application to gain global use as a means for interactively sharing content. With a "fiercely simple" design, Gopher's central server ran on twin Macintosh IIci's with twenty-five MHz processors that, even for the time, were pretty modest. Much like the Web today, emerging also in the 1990s, Gopher was a hyperlinked, multimedia, text-searchable, Web-like global file-sharing system. McCahill coined the phrase "surfing the Internet." "There is a lot to be said for . . . surfing the internet with gopher from anywhere that you can find a phone jack," he wrote in February 1992.[31] Indeed, with its search engine, you could find a photo of an obscure motorcycle on a server in Japan or, notoriously, explicit pornography on a server located on a nonterritorial island in the South Pacific beyond the reach of any laws. Building on the Internet protocols that spread worldwide in the 1980s, Gopher flourished first on the University of Minnesota campus and then, quickly, across the world. A set of annual Gopher conferences spread the word; the 1993 conference was attended by 250 invitees from Apple, IBM, Microsoft, Motorola, Xerox PARC, and the World Bank, as well as reporters from the *New York Times* and *Chronicle of Higher Education*. In May 1993 there were one thousand Gopher servers worldwide and just fifty WWW servers, while a year later there were 6,958

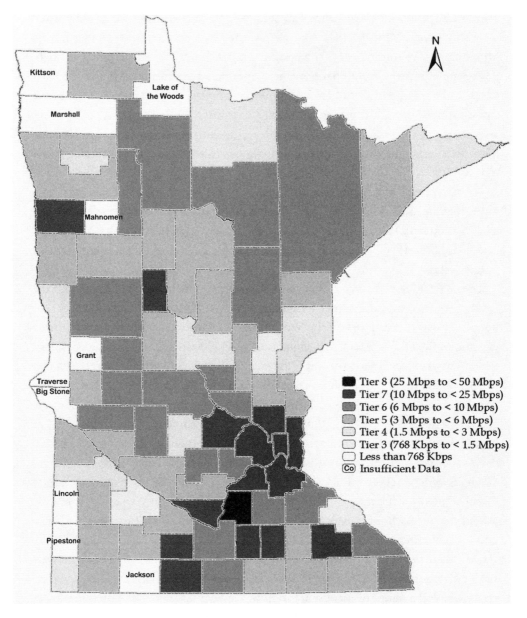

FIGURE 7.12. Internet download speeds for Minnesota, May 2012. Internet service is generally fastest in the Twin Cities metropolitan area, but Le Sueur County is a downstate hot spot. Source: Connect Minnesota, July 2012.

Gopher servers (this was 2.5 times the number of WWW servers). Gopher continued to grow, so that in spring of 1995 there were 9,046 Gopher servers, but by then the Web had exploded with more than twenty thousand servers. Internet Gopher might well have been the paradigm for the Minnesota information superhighway.[32]

Minnesota hatched at least one more grand statewide challenge, "Connecting Minnesota," that also sought to avoid a private-sector or corporate-dominated Internet. Launched in 1996 by the state's highway department (MnDOT), Connecting Minnesota aimed to build a statewide high-speed fiber-optic system using one-time access to the state's freeway right-of-ways to entice a private company to invest two hundred million dollars in laying the expensive cables. The scheme aimed directly at economic development in rural Minnesota, because (as one report put it) such "advanced telecommunications infrastructure . . . is essential to stay in the 'game' of economic development. Communities . . . may find themselves left behind, much as the railroads and the Interstate highway system left behind some communities in their early days of development."[33] Although the plan seemed promising, it was torpedoed by a lawsuit from the telecommunication industry charging that the plan was anticompetitive and monopolistic. The project missed a crucial funding deadline and, with the much-hyped dot-com bubble collapsing around it, was killed in March 2001.

As it turned out, the Internet came to Minnesota pretty much the way it did elsewhere. The University of Minnesota, and its regional counterpart MRNet, was connected through the noncommercial NSFNET (1985–95), which evolved into the high-capacity "backbone" for Internet2. Meanwhile, with the state-backed schemes pushed out of the way by lawsuits or other means, telecommunications companies flooded in to provide commercial and residential service. The telecom-based Internet was both decentralized and highly centralized. Local Internet service providers (ISPs) sprang up in many places, especially in the densely populated Twin Cities metro (see Figure 7.12), while the state's principal nonuniversity connection to the Internet writ large was, conveniently enough, in Chicago. Your e-mail to a next-door neighbor, if you had different Internet providers, went by high-capacity fiber-optic cables that ran down the state, across neighboring Wisconsin, and connected to an office building in downtown Chicago. There, Ameritech, taking control for a split second, switched your incoming e-mail over to your neighbor's Internet provider—and then it was sent back to your neighbor. Minnesota Web site content, e-commerce, and everything else on the Internet was treated much the same way. "It's pretty ridiculous that all traffic flowing between two local Minnesota service providers has to route through Chicago," noted one of the founders of MICE, the Midwest Internet Cooperative Exchange. With largely donated equipment, MICE began operating in the summer of 2011 in the 511 Building, a telecommunications hub in downtown Minneapolis.[34] At long last, Minnesota has a means for escaping the clutches of Chicago in its connection to the global Internet.

8
High-Technology Innovation

Medical Devices and Beyond

The emergence of Minnesota's industrial district, traced in this book, remains vital to the state's economy and future prospects. This chapter ties together the material in earlier chapters with a detailed assessment of the state's transition from a computer-centered industrial district to a medical-device-oriented one. It does so with firm-level data on 245 Minnesota companies active in computing across the decades spanning 1980 to 2011. The evidence demonstrates that the industrial district developed during the computing industry's glory days directly supported the state's medical-device industry, known as Minnesota's Medical Alley in a none-too-subtle parallel with California's Silicon Valley.[1]

For Minnesota's high-tech industrial district, as noted in this book's introduction, three of Phil Scranton's core concepts seem especially helpful in outlining these relationships. This book has prominently profiled the "integrated anchor" firms in computing: the small number of large firms, such as Univac, Control Data, Honeywell, and IBM that somehow managed to amass the multitudinous skills, technical competences, and organizational capabilities needed to design, build, market, sell, and service digital computers. While Univac and IBM could draw on larger national corporations, of which they were local divisions, it was something of an accident that Control Data, a start-up in 1957, was able to launch itself into the first ranks of computing and become a billion-dollar venture merely a dozen years later. In part, its early success with superfast scientific computers—in the context of the Cold War's ample military and national-security budgets—gave it the high stock-market valuation that allowed it to acquire a very large number of smaller firms over the years. In chapter 4, I stressed the acquisitions of Cedar Engineering and CEIR, key to its manufacturing and services, respectively; but it is worth noting that Control Data engaged in more than one hundred acquisitions over three decades. In 1965 alone, with its stock value riding high, it acquired an entire commercial computer division, two research houses, three computer peripherals manufacturers, two Canadian computer companies, and a Hong Kong–based manufacturer of core memory.

Far more common than the anchors, but substantially less visible, were the "specialist auxiliary" firms that manufactured components or subcomponents needed in the computer industry. These firms were literally invisible in the 1950s when computing was tallied under the Standard Industrial Classification (SIC) code 357 for business machines, but, as their numbers grew and their importance became clear, the new four-digit SIC codes spotlighted not only the core firms but also the manufacturers of computer peripherals, such as terminals or memory units. Specialized machinery suppliers and engineering design concerns, which can be located in the *Minnesota Directory of Manufacturers* (1980–2011), are another type of specialist company that was absolutely vital for the industry. My data shows that many specialized or precision manufacturers profitably rode the long wave of computing, selling products and services to the computer manufacturers for decades, and then seamlessly shifted attention to the up-and-coming medical-device industry. Local, state, or federal industrial strategies that do not recognize the vitality, and indeed centrality, of these specialist firms are missing what may be really important in industrial districts: not the high-flying anchor firms, but the low-profile networks of specialist auxiliaries.

Slightly modifying Scranton's typology, I draw explicit attention to the "ancillary industries" that form the wider ecosystem. In Minnesota, it's difficult to imagine the first generation of computer-based information processing absent the printed bank forms, bookkeeping registers, and punched data cards that created an information economy, and these items were made by companies somewhat removed from the computer industry. Similarly, as mentioned in the treatment of the pioneering Engineering Research Associates, there was a vital link to the regional giant 3M (Minnesota, Mining and Manufacturing Company), not least in the development of high-quality magnetic tapes and magnetic disks, which has its reach down to today in the 3M spin-off Imation. Also, in our age of globalization, we need to spotlight the distinctive "global enterprises" with their far-flung operations and diverse strategies. Even though some Minnesota companies own factories on every continent of the world, other Minnesota companies have been bought up by parent corporations ranging from Texas to Germany—and not always with happy results. Some global enterprises have made solid investments that have strengthened Minnesota industry; other global enterprises have taken over Minnesota enterprises, fired many workers, or even outright shut down the entire operation.

The firm-level data (fully tabulated in the Appendix) comes from carefully analyzing the thick volumes of the *Minnesota Directory of Manufacturers*. These volumes provide a unique, state-level view of Minnesota industry and—this is truly priceless—

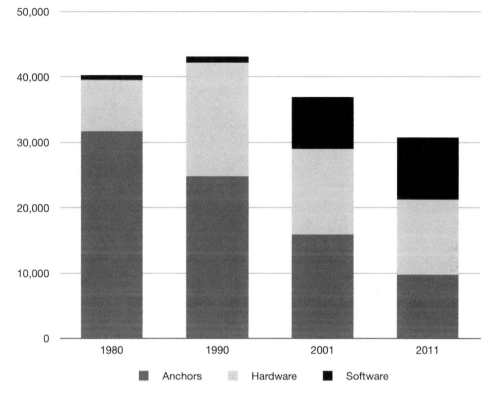

FIGURE 8.1. Minnesota employment in computing industry, 1980–2011. Source: *Minnesota Directory of Manufacturers* (years listed) tabulated in Appendix: Primary Data Software (SIC 7372) and Hardware (SIC 3571 or 3573). Anchor firms are Univac, Control Data, IBM, Honeywell, and successor companies (BT Syntegra, General Dynamics, Seagate, Unisys, Lockheed Martin).

company-by-company employment figures for the state. Its detailed listings of addresses and phone numbers allowed me to connect companies across the decades, even if their name had changed, while the indexes permitted tracking of companies as they moved around the state. I assembled a sample of 245 companies using as primary data their SIC listing in the computer hardware and software industries. Once a company was established as an anchor, specialist, or ancillary, where possible I linked the company across the decades from 1980 to 2011. Missing 2011 data from Seagate, Unisys–Roseville, and a few other companies means that employment figures for that year are very likely underestimated.

Integrated Anchor Firms

On the whole, the "integrated anchor" firms in the computing and medical-device industries have little overlap. Minnesota's core computer companies profiled in this book—Sperry Rand Univac/Unisys, Control Data, Honeywell, IBM Rochester—simply were not the spawning ground for the state's anchor firms in the medical-device industry, including Medtronic, St. Jude, Boston Scientific, and others. One important link was that technical talent flowed between firms and industries according to where the pay and prospects were good. Mike Mack was one of the numerous electrical engineers who changed industries in Minnesota without moving. "After surviving several rounds of firings at mainframe computer maker Unisys Corp., the 33-year-old electrical engineer wanted more stable work with brighter career prospects. So Mack joined Minneapolis-based Medtronic Inc., the world's largest pacemaker company," noted a newspaper account of 1997.[2] Despite retreating from its peak around 1990, the computing industry remains today, according to the Minnesota Precision Manufacturing Association, the largest in value of the state's manufactured products, followed by machinery, medical devices, and transportation equipment.[3]

The computer companies' military divisions, moreover, have provided much of the talent and staffing to the region's "systems integration" concerns, many of which have become divisions of the country's largest military contractors. The local outpost of giant General Dynamics Advanced Information Systems, in Bloomington, has its roots in Control Data's Government Systems division, which made special top-secret military-specific hardware that extended the capabilities of CDC's computers. Sperry Rand's military division, itself a descendant of the original Engineering Research Associates, emerged after several reorganizations as the Eagan division of Lockheed Martin. In 2012, Minnesota lost this part of its legacy in computing with the closing of the Lockheed Martin plant in Eagan, heir to Sperry Defense and, ultimately, heir to Engineering Research Associates.[4]

Divisions of Global Enterprise

Benchmark Electronics, now a multinational corporation that is specialized in manufacturing and engineering, came to Minnesota with its acquisition of EMD Associates in Winona. In 1980, EMD was a small firm (9–24 employees) that specialized in electronic assembly; it grew to medium size (250–499) through adding test equipment by 1990. In 1996, it was bought by Benchmark Electronics, a Texas-based medical-device manufacturer, further expanding into subcontract assembly, prototypes, and circuit boards with 950 employees in Winona by 2001. The parent Benchmark Electronics is today a global concern with nineteen design, engineering, and precision manufacturing sites in the U.S., Latin America, Europe, and Asia. Benchmark's Minnesota division—with engineering, product development, design, and manufacturing facilities in Winona, Arden Hills, and Rochester—is the company's FDA Manufacturer of Record for medical devices. It also serves the defense, aerospace, telecommunications, and networking industries. In 2011, its employment in Rochester and Winona totaled 1,800, with an additional ten employees building OEM computers in Chaska.[5]

Other Minnesota divisions of global enterprises include the high-tech anchor firms. The case of IBM Rochester (examined in chapter 6) suggests both advantages and liabilities of being a division of a global corporation. While Rochester benefited from IBM's immense resources as well as handsomely contributed its design, engineering, and manufacturing skills to IBM over many years, IBM's corporate emphasis on "computer services" in recent decades has meant that Rochester has had to scramble to maintain its relevance. Control Data, during the 1960s, became a global company with significant activity in Europe, Asia, and Australia. Univac, too, successfully sold its computing systems worldwide, including its lucrative air-traffic-control systems to European and Asian countries. Recently, the global software and services giant Tata Consultancy Services, based in Mumbai, India, opened a fifty-thousand-square-foot facility in Bloomington and is expanding its Minnesota employment to 1,300. "The decision to open a new TCS facility in the Minneapolis region is part of our company's on-going commitment to grow our presence in each and every market we serve," said the firm's chief executive officer.[6]

A number of homegrown Minnesota companies have themselves become global enterprises, such as the medical-device anchor Medtronic. Oakdale-based Imation, the well-known magnetic and optical media manufacturer, was spun off from 3M in 1996 and today has offices, laboratories, and factories in thirty-five countries around the world. More recently, the software house Open Systems, based in Shakopee and specializing in business and accounting software, opened offices in Puerto Rico, Argentina, Australia, New Zealand, and India. Hutchinson Technologies, profiled later in this chapter, has three factories in the upper Midwest and service centers in Thailand, Japan, South Korea, China, and the Netherlands. Digital River was the third Minnesota company founded by serial entrepreneur Joel Ronning, in Minnetonka, just as the World Wide Web was taking off in 1994. It successfully rode the extreme turbulence of the "dot-com" bubble that burst in 2001, lining up "outsourcing" contracts with such high-tech giants as Motorola and 3M to provide e-commerce platforms and with Symantec and Microsoft to facilitate software upgrades. If you purchased Windows Vista or Office 2007 online, you used Digital River's software. Today, Digital River, with its headquarters still in Minnetonka, employs 269 Minnesotans, around one-fifth of the company's global workforce of 1,400.[7] Along with Chanhassan-based Datalink and Mankato-based Firepond, it is one of the Minnesota companies active in creating "cloud computing."

Not all Minnesota experiences with global enterprise have been happy ones. Comten was a large manufacturer (two thousand or more employees) of communication and data computers based in Roseville. Its established expertise was in building data-communication computers that integrated with IBM networking equipment. In 1979, Comten was merged into the gigantic NCR Corporation (originally National Cash Register in Dayton, Ohio, as chapter 1 describes). Two years later, the *Computer*

Business Review noted that "NCR Corp's Comten . . . in St Paul, Minnesota seems to be run more as a secret society than as a commercial enterprise, so seldom does any news leak out of the company." Soon after being acquired by telecommunications giant AT&T, the Roseville factory disappeared from the *Minnesota Directory of Manufacturers,* and (as of this writing) the NCR–Comten Web site is, oddly enough, frozen in time in 2001.[8] Another unhappy experience with global enterprise is Telex Communication, which began as a manufacturer of hearing aids in 1936, then expanded into audio equipment and computer peripherals in the 1960s. With manufacturing plants and offices at Blue Earth, Burnsville, Glencoe, and Rochester, Telex Communication in 2001 employed nearly 1,150 Minnesotans. But in 2006 it was purchased by the giant German-based Bosch Group, and by 2011 just one in four of those Minnesota jobs were left when the dust settled.

Specialist Auxiliaries

Smaller and medium-sized "specialist auxiliary" firms are typically underappreciated but nonetheless central to the persistence of industrial districts. These smaller firms do not have the wide range of skills and organizational capabilities needed to produce complex products for sale to end users; this is the essential role of the anchor firms. Often, however, the anchor firms depend on a network of specialist auxiliary firms. You can see the clear importance of these network relationships in the metropolitan region's innumerable machine shops and engineering firms. Sometimes, as with Control Data's early acquisition of Cedar Engineering for its much-desired manufacturing capability, these firms are incorporated into the anchor firms. But often they remain as independent firms, across decades, striving to match their evolving expertise to the vicissitudes of changing markets. The pioneering ERA company's location in the Midway industrial district is a perfect example of redeploying the machine-shop skills developed for an earlier generation of manufacturing, including the nearby massive Snelling Shops complex, to manufacture subsequent generations of computers. ERA's first computers were dependent on the high-precision lathe work that created its magnetic-drum memories (see chapter 2) as well as on the better-known computer design engineers. Close examination of the *Minnesota Directory of Manufacturers* from 1980 to 2011 demonstrates that the computer sector's specialist auxiliaries—especially in design, engineering, high-precision machining, and specialist manufacturing—had a clear role in the recent development and significant expansion of the state's renowned medical-device industry. A review of specific firms can help clarify this pattern.

Bermo is a family-owned and -run specialist metalworking firm that was founded in Circle Pines (Anoka County) in 1947. Circle Pines was itself created as a 1,200-acre cooperative village in 1946, with all commercial services and municipal utilities to be owned cooperatively. The full cooperative scheme did not last long, but to this day Circle Pines has a municipally owned gas distribution system. Bermo by the 1990s was active in metal stampings, welded assemblies, injection molding, and sheet metal, with its employment fluctuating over the years between three hundred and eight hundred. Recently, the firm took up computer manufacturing, even though its principal business remains as a specialist auxiliary to Caterpillar, Polaris, and other giant manufacturing concerns (that are anchor firms in their industries). A classic "metal-bashing" firm, with its forty-five punch presses, punch breaks, and turret presses—Bermo's machine shop also features lasers and robots—it has won a reputation for high quality and quick response. Polaris, the giant Minnesota snowmobile maker, turned to Bermo in desperation for reusable snowmobile delivery crates after the original supplier flubbed the order. In just two business days, working over a weekend, Bermo's engineers and machinists designed, manufactured, and delivered the custom crates, evincing (according to Polaris) a laudable "'can do' attitude and basically flawless execution."[9]

Metalcraft Machine and Engineering in Elk River is an excellent example of a specialist auxiliary's forward linkages from machinery and computing to medical devices. The firm, established in 1978, first appeared in the 1990 *Minnesota Directory of Manufacturers* as a smaller general machining job shop (employing between twenty-five and forty-nine people) serving local markets. In 1996, Metalcraft bought Riverside Machining and Engineering in Chippewa Falls, Wisconsin (the site of Seymour Cray's supercomputing lab and factory). Metalcraft gained the Wisconsin company's expertise in the precision machining of aluminum heat sinks and heat exchangers. In 2001, the firm contributed to four SIC categories, including supercomputer hardware and medical tools, and by 2011 it had become a specialist manufacturer of medical instruments, employing ninety-eight. Among its high-precision tools, meeting a range of ISO and FDA standards, are facilities for computer numerical-controlled machining and grinding, gun drilling, and wire-cut electrical discharge machining.[10]

Hutchinson Technology illustrates the application of one firm's hyperspecialized manufacturing capability across the computing and medical-device industries. The firm, founded in the town of Hutchinson some fifty miles west of the Twin Cities in 1965, had its origins in a chicken coop rented for seventy-five dollars a month. An initial product was gyroscope heaters for missile-guidance systems. Early orders for computer circuit boards came from Univac and Control Data. It began supplying the tiny mechanisms that precisely positioned the magnetic recording heads close to the working surfaces of Control Data's disk drives, as well as making the needed electrical connections, then in 1978 gained large orders for similar parts from IBM. Lacking large-scale precision-machining capabilities itself, Hutchinson's manufacturing effort for IBM initially meant "driving to tool shops in Minneapolis." In

1980, it was a medium-sized supplier (100–249 employees) of electronic computer components, and by 1990 it grew into a large concern (2,000+) making suspension assemblies for computer hard-disk drives, now also for the burgeoning personal computer industry. It opened a second factory in Sioux Falls, South Dakota, in 1988 (with additional factories in Eau Claire, Wisconsin, and Plymouth, Minnesota, to come), as well as sales outlets in Singapore and South Korea and an overseas factory in Thailand. Its principal product, according to the company: "Suspension assemblies are precise electro-mechanical components that hold a disk drive's recording head at microscopic distances above the drive's disks." And, it further observes: "a durable competitive advantage [is] our long-standing leadership in the suspension assembly market. At one time, we had over 30 competitors [worldwide]. Today, we have only two. The others weren't able to respond to the industry's technology shifts, the intense competition and the ever-increasing market requirements." Even though computing remained its principal sector in 2001, when it employed 3,200 people, Hutchinson was reorganized into two divisions—Disk Drive Components and Bio-Measurement—with each deploying the company's expertise in ultrafine manufacturing, magnetics, and measurement. In 2011, Hutchinson employed three thousand Minnesotans, shipping a hundred million suspension assemblies worldwide in one recent three-month period.[11]

Ancillary Industries

Business forms, punched cards, circuit boards, and general-purpose machine shops constituted a set of independent "ancillary industries" that formed a base for the early computer industry. While specialist auxiliaries often developed close relations with the anchor firms and among themselves, the ancillary industries remained largely independent. Another way of expressing this point was that transactions were frequently shorter-term or market-oriented, even though there are instances of direct ties between anchors and ancillaries. Clarence Spangle had headed Honeywell's computer division since 1969 and was president of Honeywell Information Systems (1974–80), headquartered portentously enough in Waltham, Massachusetts; he then became president and CEO of the prominent magnetic-media producer Memorex. Spangle's other Minnesota connections included serving on the board of the Guthrie Theater and as trustee of the Charles Babbage Institute.[12]

In 2011, there remained a great number of Minnesota-based companies in the industries clearly ancillary to computing. These include the 20 large and small firms in communications equipment (SIC category 3669), the 29 firms in wiring devices (3643), the 46 firms in semiconductors (3674), the 54 firms in printed circuit boards (3672), and the 84 firms in electronic components (3679). In just this select sample, drawn from the electrical machinery sector, there are 233 Minnesota-based companies creating a vast industrial ecosystem for high-tech anchor firms. Many of them supplied the computer industry—for instance, in the wiring and machinery areas alone, Abelconn, Bergquist, Benchmark, Micro Dynamics, Nortech, and others.

Additional firms that serve as computing auxiliaries are the Minnesota divisions of the well-known semiconductor firms Agere Systems, Alcatel-Lucent, Cypress, Entegris, QLogic, and Texas Instruments. Silicon Valley–based Cypress Semiconductor is one of the two sizable employers (450 jobs) in that sector, with its Bloomington "fab" plant making memory, logic, and microprocessor chips. Located on Old Shakopee Road, in the shadow of the Mall of America, Polar Semiconductor (also employing 450) has its origins as a spin-off formed in 1984 from Control Data's Microcircuits Division. It was owned first by a local partnership called VTC, then by the giant Lucent-Agere (heir to Bell Laboratories), and most recently by the Japanese semiconductor manufacturer Sanken.[13] Chaska-based FSI International Inc. (three hundred jobs) makes the equipment used to manufacture semiconductor chips. Even the giant 3M concern, with its long-standing expertise in magnetic media—stretching back to the early days of the ERA company, as chapter 2 explained—was a computing "auxiliary" because it provided miles of magnetic tape for data storage. Not far from Maplewood-based 3M is its spin-off company, Oakdale-based Imation (662 jobs), which is well known for manufacturing magnetic and optical storage media.

While these figures highlight the "core" computer industry (SIC category 3571), there are, of course, a host of "peripheral" products that literally plug in to make computers useful. These include the varied products of Minnesota's twelve computer terminal makers (SIC 3575), fourteen computer-storage device makers (SIC 3572), and fifty makers of other computer peripheral equipment (SIC 3577). Some of these firms appear in the Appendix, but it is worth highlighting several prominent concerns (and their Minnesota employment in 2011): Xerox (125), which manufactures computer peripherals and telecommunication equipment in southwestern Minneapolis; Digi International (130), which makes computer peripherals in Eden Prairie; Hitachi Global Storage Technologies (160), which leases factory space inside IBM Rochester's complex; McData Services (300), which builds computer-storage devices and does computer systems design in Minneapolis; and the set of Nortech Systems plants in Blue Earth, Fairmont, Merrifield, and Wayzata (total 668).

Information services are an immense, if diffuse, industry. One can get a glimpse looking at the largest hundred publicly held companies headquartered in Minnesota.[14] By employment, Minnesota's three largest companies are mass retailers that are entirely computer-dependent—Target, Best Buy, and Supervalu, Inc., parent of Cub Foods—as are many of the other largest firms, including those in health care, finance, manufacturing, energy, and agriculture. As mentioned in chapters 4 and 6, on Control Data and IBM, a number of companies large and small have specialized in providing information and services—essentially *using* computing to deliver specialized financial, software, or management services. Some of the larger Minnesota companies specializing in information services include the Edina-based Analysts International Corp., Eden Prairie–based logistics supplier C. H. Robinson Worldwide Inc., and Minneapolis-based SPS Commerce Inc., which develops software for companies to manage their supply chains. In September 2012, Tata Consultancy Services, the immense global firm based in Mumbai, India, opened a new center in Bloomington for three hundred software and services experts, adding to its already impressive tally of one thousand Minnesota jobs.[15] The for-profit Capella Education and Dolan Media, providing business and information services to real estate, bankers, and lawyers, both in Minneapolis, are also information services companies. Then there are the communication suppliers that bring us the Internet, including Mankato-based HickoryTech Corp., New Hope–based Multiband Corp., and Hector-based Communications Systems Inc. Jay Hanke of HickoryTech took a leading role in creating MICE, the pioneering Minnesota-based Internet interchange (see chapter 7).

All of these Minnesota firms—anchor firms, specialist auxiliaries, and ancillary industries of several types—are key resources for the state's present and future high-technology industries. While much of the media and popular attention over the years inclines to the famous anchor firms, the true resilience of industrial districts, in Minnesota and elsewhere, may be in the networks of specialist auxiliaries and the wider set of ancillary industries. The hyperspecialized ultraprecision manufacturers have too frequently been overlooked, but they have been remarkably persistent. Indeed, these networks show astonishing continuity across decades, from the 1950s to the present.

The 1986 *Medical Alley Directory* presents a virtual roll call of the specialist auxiliaries that grew up around the state's computing industry. In addition to the thirty-six commercial and education concerns specializing in "computers in medicine" (and this list includes the medical-device anchor Medtronic, the University of Minnesota, and spin-offs from Control Data), there is an impressive array of manufacturing, services, and software companies with direct roots in Minnesota computing. These include numerous specialist, electronics, and high-precision manufacturing companies (Argosy Electronics, Braemar Computer, CMC Assemblers, Crest Electronics, Dahlberg Electronics, Metal Craft Machine and Engineering, Micro Dynamics, Minnesota Wire and Cable, Remmele Engineering, Waters Instruments), a set of information services companies (Control Data Healthcare Services, Qnetics, SMR Computer Services), and several application-oriented software companies (DISC Computer Systems and West Central Computers). Additional companies are clearly in the business of applying computer technologies to biomedical devices and systems. These include Aequitron Medical, Biosensor Corporation, CNS (computerized brain-wave instrumentation), Cardiac Pacemakers, Cardio-Pace Medical, Cherne Medical Group, Compucath, Degan Corporation, Data Sciences, Digimed (hospital inventory computer systems), Dimensional Medicine, Honeywell, and several dozen more.

What may be different in Minnesota's high-technology future is the diminishing position of the state's anchor firms. In the glory decades of the computer industry there were four prominent Minnesota-based anchors, but then in the medical-device industry there have been just two well-established Minnesota firms, including Medtronics and St. Jude. (Boston Scientific employs five thousand in Minnesota but is headquartered in Natick, Massachusetts.) In the more recent era of software, the most prominent local anchor firm was Lawson Software, founded in 1975 and for years a leading worldwide provider of the complex software packages that businesses use in their internal operation, competing against giants SAP and Oracle. Peter Patton, for years the University of Minnesota's director of computing, became its Chief Technology Officer, and explained one of the reasons for its success. Put simply, Lawson built an immense software firm by hiring away legions of programmers from Control Data, Unisys, and other struggling anchor firms.[16] It moved its corporate headquarters in 1999 to 380 St. Peter Street in downtown St. Paul, substantially propping up the city's economic life. In 2011, Lawson employed, according to the *Minnesota Directory of Manufacturers,* no fewer than 3,780 Minnesotans in its domestic office and an additional 900 in support of its global operations. The present and future are uncertain, however, because Lawson itself was a takeover target. As the *Star-Tribune* reported, in mid-July 2011: "Layoff notices were handed out Wednesday at St. Paul-based Lawson Software Inc., just one week after the company finalized its sale to Atlanta-based GGC Software Holdings for $2 billion."[17]

In an era of mobile and global capital, it may be impossible for a state like Minnesota to directly compete with the established global anchor firms. It was something of a miracle that Lawson lasted as long as it did. There are terrifically large sharks swimming in the software pool. New York–based Computer Associates has been notorious for snapping up any and all promising software companies—it acquired five large software houses in 2010 alone—and Oracle Corporation, the Silicon Valley giant led by Larry Ellison, continues the tradition today. Nationally, Oracle, just in the first half of 2012, made six significant acquisitions in cloud computing, social media, and enterprise software. It employs a total of 115,000 workers worldwide (around five hundred in Minnesota), roughly twice the size of Control Data at its peak of employment. The Minneapolis-based software company Retek Inc., founded in 1985, was bought by Oracle in 2005.

With each of Minnesota's anchor firms there were unusual circumstances that permitted a Minnesota-born or -grown company to develop deep pockets (that is, gain the confidence of national capital markets) and become a diversified anchor firm. All the same, anchor firms are not the only game in town. After all, the chart in Figure 8.1 indicates that Minnesota employment at the anchor firms and their successors collapsed from 1980 to 2011, falling to around one-third of the peak levels, while the networks of specialist auxiliaries and ancillary industries remain surprisingly strong. Minnesota's computer industry at the peak of the anchor firms in 1980 employed around forty thousand people, while the industry today—despite the dramatic falloff of the anchors—still employs more than thirty thousand in Minnesota. Employment in software significantly expanded (and the employment figures would be significantly higher if the companies in computer and information services, such as Tata with its one thousand-plus Minnesota jobs, could be effectively identified). It goes without saying that the full range of companies constituting a high-tech industrial district should be considered in the state's economic strategy. The diverse set of ancillary industries, given their wider base of suppliers and end markets, are often a more stable source of long-term economic activity. And the smaller and medium-sized specialist auxiliaries offer an unappreciated economic asset, which may be more flexible than the anchor firms in coping with market shifts.

Minnesota's deep wells of expertise in ultraprecision manufacturing, facility with hyperspecialized products and processes, and world-class research communities, suggest that the state is well positioned for the next generation of specialized high technology. "Over the past few years, Minnesota has put an emphasis on technological innovation that aligns with TCS' breadth of industry expertise, innovative engineering strength and commitment to providing cutting-edge solutions," stated the CEO of Tata Consultancy Services in September 2012 in announcing his company's major expansion in the metro area. Governor Mark Dayton promptly responded, "we look forward to assisting its continued success and growth in Minnesota."[18] Indeed, no fewer than seventy-two large companies located their largest R&D facility in the greater Minneapolis–St. Paul metropolitan area (which ranked sixth nationally in this measure of R&D prominence), for an annual total of nearly five billion dollars in private-sector R&D expenditures, according to a recent NSF study. Research of all types funded by the federal government accounts for $2.3 billion in the state, while the state of Minnesota itself spends a bit less than $20 million on R&D.[19]

The next generation of high-tech industry may belong to nanotechnology. One such Minnesota firm is NVE Corp., formed in the 1980s in Eden Prairie. NVE got its start with manufacturing semiconductor chips (originally "nonvolatile electronics" was its moniker), then recently has taken up "spintronics" to make an impressive entry in nanotechnology. Unlike regular electronics, where the movement of electrons alone transmits energy and information, spintronics also exploits the quantum-mechanical "spin" of electrons at the tiny scale of nanometers: "10 nanometers is 1000 times smaller than the diameter of a human hair. There are as many nanometers in an inch as there are inches in 400 miles."[20] Devices of the type NVE is manufacturing with dimensions of one nanometer, according to the citation for the 2007 Nobel Prize in physics, "can also be considered one of the first real applications of the promising field of nanotechnology."[21] NVE employs fifty Minnesotans, in one of the foremost high-tech fields in the world. Perhaps the next chapters in the ongoing history of Minnesota's high-technology industry will come out of partnerships with the University of Minnesota's Center for Nanostructure Applications in the new eighty-million-dollar Physics and Nanotechnology Building.[22] There are legacies to be remembered—and lessons to be learned—at each of these turns in the road ahead.[23]

APPENDIX
Employment in Minnesota Computing, 1980–2011

Primary data is from *Minnesota Directory of Manufacturers* SIC 3571 Electronic Computers or 3573 Electronic Computing Equipment for years tabulated. Secondary data was obtained through Hoovers profiles, LexisNexis, company Web sites, and newspapers.

For 2001–11, SIC 7372 Prepackaged Software and other software companies are denoted by bold type.

A short dash (–) indicates "no data"; a longer dash (—) indicates missing data (e.g., from a company believed to exist).

Companies with fifty or fewer employees are tabulated together at end. Companies appearing at least once in targeted SIC were linked to earlier or later years. Software companies are not always listed in MDM; see "Twin Cities Top Software Companies," bizlistr.com/company.php?id=20. Computer services are even more elusive: Tata Consultancy Services, for example, employs more than one thousand Minnesotans yet is not in this list or the MDM. Software companies are printed in bold; hardware companies are in plain text.

Company	Plant	1980	1990	2001	2011
ADP|Hollander Inc.	Plymouth 55447	–	50–74	150	130[1]
Aberon Technologies Inc.	Minneapolis 55418	–	–	–	5
Achieve Healthcare Information	**Eden Prairie 55344**	–	–	**200**	**100**
Adaytum Software Inc.	**Bloomington 55425**	–	–	90	**200**
ADC Telecommunications Inc.	Fairfax 55332– Shakopee 55379 Minneapolis 55440 **St. Paul 55110** **Eden Prairie 55344**	–	**100–249**	1175	1310
Adtech Engineering Inc.	St. Paul 55104	–	–	8	–
Advantech Inc.	Eden Prairie 55344	–	25–49	–	–
Amcom Software	**Hopkins 55343**	–	–	–	**97**
Applied Systems Inc.	**Bloomington 55425**	–	–	**125**	–
Astrocom Corporation	St. Paul 55107 Plymouth 55447	75–99	100–249	17	8[2]
Automatic Hardware Co.	Minneapolis 55413	1–8	–	–	–
BH Electronics	Burnsville 55337 Marshall 56258	–	100–249	115	110
Barclay Map Works	Shakopee 55379	–	–	–	15
Benchmark Electronics (bought EMD Associates in Winona in 1996)	Winona 55987 Rochester 55901 Chaska 55318	9–24	250–499	950	1810

Bergquist Switch	Edina 55435	—	—
Bergquist Co. Inc.	Cannon Falls 55009	—	158
Ball Electronic | Bermo Inc.	Circle Pines 55014	250–499[3]	850	300
Brom Microsystems Engineering	Winona 55987	1–8	—	—
Business Improvement BI Worldwide	**Edina 55439**	—	**1400**	—
ByteSpeed LLC	Moorhead 56560	—	—	45
CA | Computer Associates Inc.	**Minneapolis 55435**	—	**70**	**170**
Color-Ware Inc.	Winona 55987	—	—	17
Comped	Belgrade 56312	—	—	5
Compellent Technologies	Eden Prairie 55344	—	—	142
Compudyne	Hibbing 55746	—	23	12
Compudyne	Duluth 55802	—	—	8
Computer Designed System Inc.	Minneapolis 55441	50–74	—	—
Computer Network Technology | CNT	Maple Grove 55369 Plymouth 55442	75–99	300	1000[4]
Computers by Design of NM	Minneapolis 55449	—	—	3
Comten | NCR-Comten (merged in 1979; later purchased by AT&T)	Roseville 55113	1000–1999	2000+	—
Control Data[5]	50+ Minnesota plants	19,178[6]	see note[7]	—
Corporate Travel Services	**St. Paul 55101**	—	—	**75**
Cosmic Cowboy	Janesville 56048	—	—	1
Cray Research Inc. | Silicon Graphics (Eagan)	Mendota Heights 55120	250–499	750–999	600[8]	225[9]

Company	Plant	1980	1990	2001	2011
Creative Vision Technologies	Hamel 55340 Plymouth 55441	–	–	9	15
CTS Corporation | AbelConn | (Minneapolis)	New Hope 55428	(Fabritek)	2000+	200	150
D&B Computer Systems	Hugo 55038	–	–	–	6
Data Card Corporation, Minneapolis	Minnetonka 55343	500–749	2000+	1000	800
Data 100 Corporation (to Northern Telecom)	Minnetonka 55343	2000+	–	–	–
DataSource Pomeroy	**St. Louis Park 55416**	–	–	**60**	**60**[10]
Dicomed Corporation | Crosfield	Minneapolis 55431 Burnsville 55337	25–49	100–249	30[11]	–
Di-Hed Yokes Manufacturing	Minneapolis 55406 Lakeville 55044	–	25–49	60	50
Digital Playhouse Inc.	Bethel 55005	–	–	–	1
Digital River	**Eden Prairie 55344**	–	–	–	**269**
DSC Nortech Electronics Nortech Systems	Bemidji 56601 Wayzata 55391 Blue Earth 56013	–	100–249	300	150
					215
Earth First Computer Recycle	Anoka 55303	–	–	–	2
EMC Corporation	**St. Paul 55101** **Eden Prairie 55344**	**50–74**	**100–249**	**80**	**120**
ENetpc	Eden Prairie	–	–	18	–

APPENDIX 237

Company	Location	Size range		
Epicor Software	St. Louis Park 55426	—	—	140	125
	Minneapolis 55426			
Equus Computer Systems	Minneapolis 55414	—	250	100
	Minnetonka 55343			
Fabrico Inc.	Minneapolis 55428	9–24	—	—
Fabritek Inc. (bought by CTS, Elkhart, Indiana)	Minneapolis 55436	1000–1999	(see CTS)	—
Fair Isaac Corp.	**Minneapolis 55402**	—	—	**322**
Fieldworks Inc.	Eden Prairie 55344	—	70	—
Firepond Inc.	**Mankato 56001**	—	**250**	**40**
Fpx LLC	**Mankato 56001**	—	—	**58**
General Business Systems	Eden Prairie 55344	9–24	—	—
General Dynamics Information	Bloomington 55431 55425	—	915	632[12]
General Fabrication Corporation	Pine City 55063	50–74	—	—
George Konik Associates	**Minneapolis 55439**	—	—	**70**
H D Hunter & Associates	Minneapolis 55408	9–24	—	—
Healy-Ruff Inc., Minneapolis	St. Paul 55113	25–49	25	1
Honeywell Research|Technology	Minneapolis 55418	250–499	400	500
Hutchinson Technology	Hutchinson 55350	2000+	3200	3000
	Plymouth 55442			
Imation Corp.	Pine City 55063	—	542	662
	Oakdale 55128			

Company	Plant	1980	1990	2001	2011
Innovative Computer Concepts	Hibbing 55746	–	–	–	6
Insignia Systems Inc.	**Plymouth 55446**	–	–	70	–
Instrumentation Services Inc.	Minneapolis 55427	25–49	25–49	13[13]	–
Integrated Systems, Inc.	Minneapolis 55441	–	–	–	6
Intograms (division of Interactive Software)	**Plymouth 55447**	–	–	**80**	–
IBM	Rochester 55901	6000[14]	8100[15]	6000	5100
Kewill ERP	**Edina 55435**	–	–	**100**	**< 5**
Kleffman Electronics Inc. (KEI)	Minnetonka 55343	9–24	–	–	–
Kroll Ontrack Inc.	**Eden Prairie 55347**	–	–	–	**210**
KS/CT Manufacturing Inc.	New Ulm 56073	–	9–24	–	–
Kurt Manufacturing Company	Blaine 55449 Minneapolis 55421	100–249	1000–1999	793	431
Lakeview Technology Inc.	**Rochester 55901**	–	–	55	77[16]
Lawson Software	**St. Paul 55102**	100[17]	250[18]	1500	900
Letourneau Roy G Company	Minneapolis 55418	1–8	–	–	–
Link Datacom Company	Byron 55920	–	–	–	23
Lucent Technologies	Alcatel	Eden Prairie 55344 Elk River 55330	–	–	60	8
MWD Research Inc.	Minneapolis 55418	1–8	–	–	–
Makemusic Inc.	**Eden Prairie 55344**	–	–	–	**54**

Marner Micro Technologies Inc.	Stillwater 55082	—	—	20	24
Maxwell Technologies Inc.	Minneapolis 55421	—	—	—	35
Metalcraft Machine/Engineering	Elk River 55330	—	25–49	51	98
Micro Dynamics Corporation	Montevideo 56265	—	50–74[19]	150	125
Microsoft Corporation	**Minneapolis 55437**	—	—	—	**80**
MQSoftware Inc.	**Minneapolis 55416**	—	—	—	**100**
MNC International	Minneapolis 55418	—	9–24	3	—
Mag-Tech Inc.	Brooklyn Park 55429	—	9–24	—	—
Micro Display Systems Inc. Genius Technologies	Hastings 55033	—	25–49	—	—
MJS Systems	Eagan 55121	—	9–24	—	—
National Computer Systems	**Owatonna 55060**	**500–749**	**200–499**	**300**	**600**
NCS Pearson Inc.	**Minneapolis 55435**				
	Eden Prairie 55344				
	Minneapolis 55437				
Network Systems Corporation (merged in 1995 with Storage Technology Corporation; bought by Sun in 2005)	Brooklyn Center 55430 Minneapolis 55428	100–249	250–499	—	—
Net Perceptions	**Edina 55435**	—	—	150	—
Next Century Technology Inc.	Burnsville 55337	—	—	10	—
Nonvolatile Electronics (NVE)	Eden Prairie 55344	—	—	63	50
Northern Telecom Inc.	Minnetonka 55343	—	250–499	—	—
Nternational Products	**Minneapolis 55426**	—	—	—	**58**
Object FX Corporation	**St. Paul 55114**	—	—	60	27

240 APPENDIX

Company	Plant	1980	1990	2001	2011
Omnium Corporation	Stillwater 55082	—	1–8	—	—
Open Systems Inc.	**Shakopee 55379**	—	—	—	**100**
Oracle America Inc.	Minneapolis 55435	—	—	—	65
Oracle Corporation	Bloomington 55437	—	—	280	**500**
	Minneapolis 55403	—	—	—	—
P C Tech	Lake City 55041	—	9–24	—	—
Paisley Consulting Inc.	**Cokato 55321**	—	—	33	**125**
Patterson Co. Inc.	**St. Paul 55120**	—	—	—	**185**
Plato Learning	**Bloomington 55437**	—	—	—	**170**
Portico Computers II Inc.	Bloomington 55420	—	—	15	—
Precision Diversified Industries	Plymouth 55441	—	75–99	—	—
Print Cafe/Hagen	Eden Prairie 55344	—	—	140	—
Publishing Business Systems	**Roseville 55113**	—	—	100	**90**
Rackmaster Systems Inc.	Bloomington 55420	—	—	16	8
	Shakopee 55379	—	—	—	—
Reason Computer Inc.	Lino Lakes 55014	—	—	17	—
Research Inc.	Eden Prairie 55343	100–249	250–499	130	129[20]
Response	Rochester 55906	—	—	55	—
Retek Inc. (founded 1985; Oracle bought 2005)	**Minneapolis 55402**	—	—	**461**	—

APPENDIX 241

Company	Location				
Rimage Corporation	Eden Prairie 55346 Edina 55439	—	100	—	123
Rockwell Automation | DataMyte	Minnetonka 55345 St. Paul 55113	25–49[21]	100–249	100	60
SGI (Silicon Graphics)	**St. Paul 55121**	—	—	**400**	—
SPSS (ShowCase) (IBM bought in 2009)	Rochester 55901	—	—	—	**100**
Scantron Corporation	**Eagan 55121**	—	—	—	**200**
Seagate Technology	Minneapolis 55440 Shakopee 55379	—	4,500[22]	4000[23]	1500[24]
Select Sales Inc.	Minneapolis 55439	—	—	—	33
Shafer Electronics	Shafer 55074	—	100–249	150	150[25]
Shavlik Technologies Corporation	**St. Paul 55113**	—	—	—	**95**
Siemens Product Lifecycle Management	**St. Paul 55126**	—	—	—	**100**
Silacon Corporation	Woodbury 55125	—	—	—	3
Sisu Medical Solutions LLC	**Duluth 55802**	—	—	—	**75**
Sperry Univac | Unisys	St. Paul 55165	2000+	2000+	—	20
Sperry Univac | Unisys (ERA plant)	St. Paul 55165	500–749	—	—	—
Sperry Univac | Unisys (ERA plant)	Roseville 55113	2000+	2000+	1000[26]	—
Sperry Univac Lockheed Martin	Eagan 55121	1000–1999	2000+	2500	1600
Standard Register Company	Duluth 55802	—	2000+	—	41
Sungard Financial Systems	**St. Paul 55121, 55101 Hopkins 55343**	—	—	—	**290**

Company	Plant	1980	1990	2001	2011
Symantec Corporation	**St. Paul 55113**	–	–	–	**500**
Syntegra[27] | BT Syntegra	**St. Paul 55126**	–	–	1100	**400[28]**
Technology 80 Inc.	Minneapolis 55427 Maple Grove 55369 Eden Prairie 55344	–	9–24	27	4
Telex Communication Inc.[29]	Blue Earth 56013	250–499	250–499	275	–
Telex Communication Inc.	Burnsville 55337	–	–	270	283
Telex Communication Inc.	Glencoe 55336	–	–	300	–
Telex Communication Inc.	Rochester 55903	–	–	300	–
Tempworks Software	**St. Paul 55120**	–	–	–	**54**
Ternes Corporation	St. Paul 55113	–	25–49	–	–
TestQuest	**Eden Prairie 55344**	–	–	62	**40**
Tori Corporation	Minneapolis 55435	–	1–8	–	–
Tricord Systems Inc.	Minneapolis 55441	–	–	–	37
ViA Inc.	Burnsville 55337	–	–	50	–
Volks Communications Inc.	Long Lake 55356	–	–	–	1
Virtelligence Inc.	**Eden Prairie 55346**	–	–	–	**107**
Vital Images Inc.	**Minnetonka 55343**	–	–	–	**105**
Vomela Specialty Inc.	St. Paul 55101	25–49	75–99	65	136
Winland Electronics Inc.	Mankato 56001	–	25–49	140	82

Winnebago Software (merged in 2000 with Sagebrush, Burnsville)	Caledonia 55921	—	300	80[30]
XATA Corporation	Burnsville 55337 Eden Prairie 55344	—	60	115
X-Late Inc.	Lakeland 55043	—	—	6
Xerox Corporation	Edina 55435 Minneapolis 55431	100–249	350	140
Xpress Systems Inc.	New Prague 56071	—	—	5
Zytec | Artesyn Technologies	Redwood Falls 56283 Eden Prairie 55344	500–749	550	102
Sum of 85 smaller software companies (less than fifty employees)	various locations	—	387	1604

Notes

Preface

1. Owens, "Vannevar Bush and the Differential Analyzer," 63n2.

2. Philadelphia-centered accounts of early computing include Stern, *From ENIAC to UNIVAC;* McCartney, *ENIAC.*

3. See Westwick, *Blue Sky Metropolis.*

4. Tomash and Cohen, "The Birth of an ERA." A valuable source on these early years is Lundstrom, *A Few Good Men from Univac.*

5. Murray, *The Supermen;* Worthy, *William C. Norris.* A valuable essay on Cray as a "charismatic engineer" is in MacKenzie's *Knowing Machines,* chapter 6. Control Data of Canada figured in Vardalas, *The Computer Revolution in Canada.*

6. Price, *The Eye for Innovation;* Misa, ed., *Building the Control Data Legacy.*

Introduction

1. The "Supercomputer Capital of the World" formed a mock oversize license plate issued as a press release by Governor Rudy Perpich on November 21, 1983. Original held by CBI.

2. For the state's early history, with full treatment of the fractious interactions between French traders, British colonists, and the several tribes of Native Americans, culminating in the 1862 Dakota massacre, see Wingerd, *North Country.*

3. Lewis, *Babbitt* (1920), chapter 7, available at tinyurl.com/8oqynrc. Sauk Centre is identified as a model for Lewis's *Main Street* (1920). (All urls cited in these notes as tinyurl.com are expanded at www.cbi.umn.edu/digitalstate/.)

4. Norberg, *Computers and Commerce.*

5. Ralph Mason, "Twin Cities Boom: Electronics: Where 'Little Guys' Get Big," *Minneapolis Star* (April 2, 1959); "General Mills: A Small Computer Operation," *Upper Midwest Investor* 1:7 (November 1961): 17–18.

6. Mason, "Twin Cities Boom." The persisting concentration of the computing industry is a puzzle for locational analysis. Flawed data on Minnesota mars the analysis of Beardsell and Vernon Henderson, "Spatial Evolution of the Computer Industry in the USA."

7. Special issue, "The Computer Industry," *Upper Midwest Investor* 1:7 (November 1961): 4 (computers were a natural), 32 (few metropolitan areas), 41 (eight thousand people). On weather, see Rappaport, "Moving to Nice Weather," 375–98. "Duluth Minnesota's weather *without* air conditioning is preferred to that of numerous cities in the South and Southwest. But *with* AC, Duluth's weather is the least preferred among the depicted cities" (379; emphasis added). Minneapolis is similar to Duluth, whereas such latter-day high-tech hotspots as San Francisco, San Jose, Seattle, and Portland rank among the cities with the most desired weather (ibid.).

8. David Barboza, "In Roaring China, Sweaters Are West of Socks City," *New York Times* (December 24, 2004), available at tinyurl.com/6b3tg (November 2011).

9. Marshall, *Principles of Economics,* book 4, chapter 10, § 3, available at tinyurl.com/j2ab8 (September 2012).

10. Scranton, *Endless Novelty.* See also Belussi and Caldari, "At the Origin of the Industrial District," 335–55; Markusen, "Sticky Places in Slippery Space," 293–313; Zeitlin, "Industrial Districts and Regional Clusters," 219–43.

11. Elizabeth Starling, "The Financial Services Cluster of the Twin Cities," Metropolitan Council (October 1995); and Laura Bolstad, Dan Maloney, and Cynthia Yuen, "The Financial Services Cluster of the Twin Cities," University of Minnesota Humphrey School of Public Affairs (May 4, 2010), available at tinyurl.com/9nbrnxs (September 2012).

12. Lécuyer, *Making Silicon Valley;* Ceruzzi, *Internet Alley.* Other recent accounts include Berlin, *The Man behind the Microchip;* House and Price, *The HP Phenomenon.* A classic is Saxenian, *Regional Advantage.*

13. Milward, *War, Economy, and Society, 1939–1945;* Smith, ed., *Military Enterprise and Technological Change;* Hooks, *Forging the Military–Industrial Complex.*

14. As detailed below, the British code-breaking efforts at Bletchley Park were kept entirely out of the public eye until the 1970s, when the remarkable early computer known as Colossus was finally made public. See Winterbotham, *The Ultra Secret;* Copeland et al., *Colossus.*

15. For von Neumann's notable career, see Aspray, *John von Neumann and the Origins of Modern Computing;* Heims, *John von Neumann and Norbert Wiener.* A Neumann-centric view colors Dyson, *Turing's Cathedral.*

16. Robert Emmett McDonald oral history, CBI OH 57, 16–17 (not online).

17. By comparison, Boston and the area south of San Francisco that in time became Silicon Valley, two other leading districts that later developed computing expertise, were each around 1960 firmly specialized in electronic *components,* such as transistors and early integrated circuits. DEC was founded as Digital Equipment Corporation in 1957, with a distinct aversion to "computing," while Fairchild Semiconductor, also founded in 1957, which spun off the electronics giant Intel, was likewise a components company, not a computing one. "Silicon Valley" was itself named only in 1971.

18. Lampe, *The Massachusetts Miracle;* Saxenian, *Regional Advantage.*

19. See Silicon Valley Historical Association, "Wagon Wheel Restaurant," tinyurl.com/79fh2dn (May 2012).

20. Evaluations of the legal decision are sharply polarized. The Philadelphia-inclined followers of J. Presper Eckert and John Mauchly have generally condemned the decision that invalidated the ENIAC patent. In contrast, the followers of John Atanasoff, named there as the true inventor of the computer, have strongly praised the decision; see, e.g., Burks, *Who Invented the Computer?*. Recently, in a book that has not been well received by historians, novelist Jane Smiley has written a thinly sourced account supporting Atanasoff: *The Man Who Invented the Computer*.

21. See tinyurl.com/7c9jsxd (September 2012). Seymour Cray was the only other computer figure so named.

1. Philadelphia Story

1. Agar, *The Government Machine*.
2. Misa and Seidel, eds., *College of Science and Engineering*, 103.
3. Kahn, *Codebreakers*, 744. See also Petzold, *The Annotated Turing*; I. B. Cohen and Owen Gingrich, "Notes on an Interview with Claude Shannon," May 27, 1970, I. B. Cohen papers CBI (182), box 3, folder Claude Shannon interview notes; Hodges, *Alan Turing*, 245–55.
4. Reynolds, *In Command of History*, 161–63, 184–85, 305, 413, 502, 537; Copeland et al., *Colossus*, 2–3, 101–15, 172–75, 295.
5. Layton, *And I Was There*, 285.
6. Ibid., 286–304. See "Joint Committee on the Investigation of the Pearl Harbor Attack: Appendix D," 434 (FDR and war), at tinyurl.com/3tfylpw.
7. Van Der Rhoer, *Deadly Magic*, 48–52, 60–66.
8. Quote from Naval Security Group Station History (July 13, 2008) at tinyurl.com/5uycyz.
9. Safford and Wenger, *U.S. Naval Communications Intelligence Activities*, 19–57; DeBrosse and Burke, *The Secret in Building 26*, 56–58, 65–66 (on Wenger); H. Dick Clover, CBI OH 113, 19 (on Engstrom); Hugh Duncan, CBI OH 118, 8 (Norris during war). On Driscoll, see NSA biography at tinyurl.com/3f9lbgj (June 2011); Howard Campaigne (33–36 [on Driscoll], 5–7, 14, 39–41, 120–22 [on Wenger]) in his secret June 28, 1983 interview, now declassified, available at tinyurl.com/6d7dqgg (June 2011).
10. H. Dick Clover, oral history interview, Charles Babbage Institute OH 113, 21.
11. James T. Pendergrass oral history, Charles Babbage Institute OH 93, 5.
12. DeBrosse and Burke, *The Secret in Building 26*, 107 (child of IBM); Van Der Rhoer, *Deadly Magic*, 83 (go on working), 70–103 (Midway code breaking).
13. Layton, *And I Was There*; Safford and Wenger, *U.S. Naval Communications Intelligence Activities*, 12–14; Parker, *A Priceless Advantage*.
14. More precisely, the Lorenz encrypting scheme logically "added" a dynamically generated stream of letters (a key) to the letters of the original message's plaintext, generating an enciphered Morse-code-like string that was directly sent by radio. At the receiving station, the decrypting machine logically "added" the same string of key-letters, producing the original plaintext (for example, where A is plaintext, N is key-text added by the machine, and Q is the transmitted letter: A + N = Q but also Q + N = A). The twin wheels that generated the key-stream of letters were synchronized in the sending and receiving machines.
15. Copeland et al., *Colossus*, 32–35, 37–51, 206–7, 378–83.

16. H. Dick Clover, oral history interview, Charles Babbage Institute OH 113, 31 (on Meader); "History of Op-20-G-4E," at tinyurl.com/3hj68ho (April 2011).

17. Turing quote from "History of the Bombe Project" (April 1944) at tinyurl.com/43at mk6 (March 2011). It is possible that some details of the British bombe were transferred when Joe Eachus, whom Engstrom had sent to Bletchley Park, went to Dayton in May 1943; see Burke, *Information and Secrecy*, 283–303, esp. 295, 431n27. Desch is the central figure in DeBrosse and Burke, *The Secret in Building 26*, 10–11, 73–77 (blueprints), 174, 194 (MIT and IBM).

18. Robert E. Mumma, oral history interview, Charles Babbage Institute OH 73, 13–14 (not at liberty, hush-hush); Carl F. Rench, oral history interview, Charles Babbage Institute OH 72, 10–13.

19. Lee, Burke, and Anderson, "The US Bombes, NCR, Joseph Desch, and 600 WAVES," 35; DeBrosse and Burke, *The Secret in Building 26*, 127–45.

20. DeBrosse and Burke, *The Secret in Building 26*, 139.

21. James T. Pendergrass, oral history interview, Charles Babbage Institute OH 93, 5 (sunk sub), 6 (wheel numbers); DeBrosse and Burke, *The Secret in Building 26*, 144 (knock on door).

22. A version is in Halmos, "The Legend of John von Neumann," 386–87.

23. Owens, "Vannevar Bush and the Differential Analyzer," 63–95; Lemaitre and Vallarta, "On the Geomagnetic Analysis of Cosmic Radiation," 719–26. More than thirty papers in *Physical Review* during 1931–51 utilized the MIT differential analyzer in calculating results.

24. Kathleen McNulty, in Shurkin, *Engines of the Mind*, 127–28; Light, "When Computers Were Women," 455–83.

25. Carl Chambers, oral history interview, Charles Babbage Institute OH 7, 4.

26. Quote from William T. Moye, ARL Historian, "ENIAC: The Army-Sponsored Revolution" (January 1996), at tinyurl.com/3nbgdbh (February 2011).

27. "It wasn't even a draft when he [von Neumann] wrote it. He wrote these as letters to Goldstine, and when we asked what he was doing this for at the time, Goldstine said, 'He's just trying to get these things clear in his own mind and he's done it by writing me letters so that we can write back if he hasn't understood it properly.' That's the basis in which he wrote it," is Eckert's post-facto recollection. See J. Presper Eckert, oral history interview, Charles Babbage Institute OH 13, 35.

28. Fritz, "The Women of ENIAC," 20.

29. Goldstine, *The Computer from Pascal to von Neumann*, 229.

30. Campbell-Kelly and Williams, eds., *The Moore School Lectures*, xv–xix; Frank Verzuh notes, CBI (51), folder 2, 205 (confident), 210 (plan and manner).

31. Louis D. Wilson in UNIVAC Conference, CBI OH 200, 134. Isaac L. Auerbach oral history, CBI OH 2, 8, 11–12, confirms Eckert and Mauchley's wild pricing on early UNIVAC contracts: "I stated that in my opinion neither Eckert nor Mauchly were competent businessmen. I had been brought in to try to negotiate the UNIVAC contract, and one of the questions was its price. Eckert gave me his costs which were $80,000 to build a computer. A UNIVAC—UNIVAC I. And I came back and said 'there's no possibility of our doing it for that price.' Just taking the bill of materials that Eckert gave me, and knowing what little I did at that time, I was to come up with the cost of the UNIVAC in two days. I had a number that was like $180,000 or $190,000. Eckert said categorically I didn't know what the hell I was

talking about, I didn't know anything about electronics; but he did, and that he would set the price. He made a major concession by raising the price 20% to $96,000 which, as I recall, was the face value of the contract."

32. Stern, *From ENIAC to UNIVAC*, 117, 124.

33. In his deposition for the *Honeywell v. Sperry Rand* case, Mauchly specifically mentions the NSA product twice but does not describe it; see Honeywell vs. Sperry Rand Records, 1846-1973, CBI (1), box 47, folder 6, 41–42. Isaac Auerbach stated emphatically that, while he worked for the Eckert–Mauchly company (from April–May 1947 through early 1949), "There were no classified contracts whatsoever in the company when I joined it, or during the entire tenure of my stay at the company" (Isaac L. Auerbach oral history, CBI OH 2, 4).

34. Stern, *From ENIAC to UNIVAC*, 149.

35. Not everyone was positive: when "Remington Rand bought [the Eckert-Mauchly company], things changed completely. Absolutely. The sales people got in, and the whole thing just changed. Women, as far as I could see, had absolutely no future under Remington Rand, absolutely none," according to Frances E. Holberton oral history, CBI OH 50, 12.

36. H. Dick Clover, oral history interview, Charles Babbage Institute OH 113, 20.

37. "Wenger felt that he was afraid that all this talent he had collected would be dissipated. And he came up with the idea of having them form an independent, private organization which would keep them together. And if the Navy would support them, the Navy would help them stay together and they'd be available there anytime an emergency arose. And they talked about this idea to a considerable extent and they organized what eventually they called the Engineering Research Associates. And Engstrom was one of the prime movers in that, a central figure. And Bill Norris was also a prime mover. He was quite energetic and pushing the idea," according to Howard Campaigne in his secret June 28, 1983, interview, now declassified, available at tinyurl.com/6d7dqgg (39) (June 2011).

38. OP-20-G, "Secret: Memo for OP-20: Research and Development Plan," February 22, 1945 in CBI (100), box 1, folder 9.

39. James T. Pendergrass oral history, Charles Babbage Institute OH 93, 10.

40. Robert E. Mumma, oral history interview, Charles Babbage Institute OH 73, 16.

41. Willis K. Drake, oral history interview, Charles Babbage Institute OH 46, 21.

42. John E. Parker, oral history interview, Charles Babbage Institute OH 99, 16.

43. Northwestern Aeronautical Corporation plant brochure [n.d.] in CBI (100), box 1, folder 8; H. Dick Clover, oral history interview, Charles Babbage Institute OH 113, 16–17 (CSAW and Forrestal); John Parker to Francis J. McNamara, Remington Rand general counsel, January 12, 1952, in CBI (100), box 1, folder 6 (on ERA financing); John E. Parker, oral history interview, Charles Babbage Institute OH 99, 23.

44. "The Bureau of Ships contract . . . was transferred from NAC to ERA . . . early in August 1947. The task numbers looked entirely different from those in the early BuShips contract. The GOLDBERG project was listed as Task 1H on the old contract, but GOLDBERG turned out to be Task 9 on the new one" (Arnold A. Cohen oral history interview, Charles Babbage Institute OH 58, 27).

45. At least one early employee of ERA, William R. Boenning, returned shortly to join what became NSA. As Cohen recalled, Boenning "went very early in the game, perhaps 1947, late 1947, to join NSA, or whatever it was known as at that time. And he was there until

retirement. Rose into middle management of that agency" (Arnold A. Cohen, oral history interview, Charles Babbage Institute OH 138, 7).

46. George Champline, in George Gray, "Engineering Research Associates and the Atlas Computer," *Unisys History Newsletter* (June 1999), at tinyurl.com/4xuuuvy (March 2011).

47. John L. Hill, oral history interview, Charles Babbage Institute OH 101, 55.

2. St. Paul Start-up

1. For the history of St. Paul's Midway district, see www.hamlinemidwayhistory.org and the sources cited below.

2. Minnesota Federal Writers' Project, *The WPA Guide to Minnesota*, 164–65.

3. Diers and Isaacs, *Twin Cities by Trolley*, 65–69, 128–29.

4. Millett, *AIA Guide to the Twin Cities*, 502; Minnesota Historical Society photos of the Griggs–Midway building, including inside views of the canning factory, are at tinyurl.com/6akbbbs (June 2011); Leacock collection of ERA and Univac scrapbooks CBI (57).

5. Willis K. Drake oral history, CBI OH 46, 5 (beat-up chair); Walter L. Anderson oral history, CBI OH 119, 5 (little wings).

6. Walter L. Anderson oral history, CBI OH 119, 13 (railroad rooms); engineering manager Lowell Benson comment (special clearance), vipclubmn.org/facilities.aspx (March 2011).

7. George Champine, quoted in George Gray, "Engineering Research Associates and the Atlas Computer (UNIVAC 1101)," *Unisys History Newsletter* 3:3 (June 1999), at tinyurl.com/4xuuuvy (March 2011).

8. Hugh Duncan, CBI OH 118, 14.

9. Robert Emmett McDonald oral history, CBI OH 45, 16.

10. H. Dick Clover oral history, CBI OH 113, 27. On NSA, see Burns, "The Origins of the National Security Agency 1940–1952." Years later, Frank Mullaney repeatedly used "the customer" to *avoid* speaking NSA's name in his CBI oral history; see the same circumlocution in W. G. Welchman to H. T. Engstrom, October 17, 1952, CBI (176), box 6, folder 13.

11. Arnold A. Cohen oral history, CBI OH 58, 27, 30.

12. In *Information and Secrecy*, Colin Burke traces the varied strands of plans and activities that connected Bush, wartime code breaking, and the postwar efforts of CSAW–NSA and the ERA company, while also making it clear that these strands did not neatly come together as a well-braided rope.

13. NCML, remaining a separate naval unit, inspected and certified ERA's early research work and technical capacities. "In August 1947 [NCML personnel] vouched that ERA now had the technical skill to design and build the large digital computer" that became Atlas, and so "at this point BUSHIPS . . . awarded a new contract directly to ERA. Task 13 . . . directed design of a complete general-purpose, stored-program digital computer," according to Boslaugh, *When Computers Went to Sea*, 93.

14. Walter L. Anderson oral history, CBI OH 119, 7 (frequency counts); Frank C. Mullaney oral history, CBI OH 110, 20 (whole matrix), 22 (make up things); Arnold A. Cohen oral history, CBI OH 58, 22 (sixty mph).

15. John L. Hill oral history, CBI OH 101, 43 (Demon as killer); Frank C. Mullaney oral history, CBI OH 110, 21 (lot of elements); Mullaney recalled: "I spent quite a bit of time in Washington [at Nebraska Avenue with the customer, i.e., NSA] nursing that machine [Demon] after we delivered it, too" (31).

16. James T. Pendergrass oral history, CBI OH 93, 20.

17. NSA, "Before Supercomputers: NSA and Computer Development" (DOC ID: 3575750), originally "top secret" but declassified 2003, available at tinyurl.com/cbfdgzc (July 2012). For Wikipedia, see tinyurl.com/c57czru (April 2013).

18. John L. Hill oral history, CBI OH 101, 45; Norberg, *Computers and Commerce*, 120.

19. Engineering Research Associates, *High-Speed Computing Devices* (1950), 301; Brusentsov and Alvarez, "Ternary Computers," 74–80. The "optimum" base for computing is the natural number e, or 2.718, according to Hayes, "Third Base," 490.

20. Arnold A. Cohen oral history, CBI OH 138, 25–26. See also Burke, "An Introduction to a Historic Computer Document," 65–75, at tinyurl.com/3vt659 (May 2011). In his oral history, Mullaney mentions technical interactions on Atlas also with Joe Eachus.

21. With Atlas I, it seems likely that James Pendergrass and, possibly, Joe Eachus were living and working in St. Paul. With Atlas II, "some of the customer's personnel lived in St. Paul for a considerable period of time, so we were with them every day. They were in the lab watching us work; they were helping us test the machines and so forth to get acquainted with them," recalled Frank C. Mullaney oral history, CBI OH 110, 58.

22. James T. Pendergrass oral history, CBI OH 93, 19; Arnold A. Cohen oral history, CBI OH 138, 27; Snyder, *History of NSA General-Purpose Electronic Digital Computers*, 10, 13, 96, at tinyurl.com/265zvjn (June 2011). ERA counterparts at NSA were Howard Campaigne (formerly a math professor at the University of Minnesota), Joe Eachus, and James Pendergrass. Atlas instruction 15 (misleadingly titled "vector add") made it a classified cryptography machine. The *non-carrying* addition was useful in stripping out the "additive" codes used, for example, by the Japanese.

23. Arnold Cohen notes (October 14, 1947), in CBI (176), box 6, folder 10 (Cognac); Arnold A. Cohen oral history, CBI OH 58, 8 (scaling counters).

24. Frank C. Mullaney oral history, CBI OH 110, 17 (pulses), 30 (machine instructions); John L. Hill oral history, CBI OH 101, 45, 47, 48. It was CSAW–NSA that directed ERA "to use whatever was available in the way of designs and concepts at other facilities to accomplish faster their navy objectives on the new Atlas being designed," according to Norberg (*Computers and Commerce*, 122).

25. Sidney Michel Rubens oral history, CBI OH 100, 55–59, 76–77, 56.

26. Frank C. Mullaney oral history, CBI OH 110, 26; John L. Hill oral history, CBI OH 101, 49, 77 (current density and submarine hatch); Boslaugh, *When Computers Went to Sea*, 92–93, 96 (inspectors).

27. Frank C. Mullaney oral history, CBI OH 110, 23; John L. Hill oral history, CBI OH 101, 51, 53.

28. Frank C. Mullaney oral history, CBI OH 110, 41.

29. Years later, Remington Rand disclosed that the sales prices quoted to NSA for two copies of Atlas I were $997,808 for the first and $287,600 for the second copy; see Norberg, *Computers and Commerce*, 154. The first Atlas I "cost about $950,000," according to Snyder, *History of NSA General-Purpose Electronic Digital Computers*, 8, at tinyurl.com/265zvjn (June 2011). The "approximate price for the basic system" for a UNIVAC was $930,000, according to Weik, *A Survey of Domestic Electronic Digital Computing Systems*, 180.

30. Stern, *From ENIAC to UNIVAC*, 127; Isaac L. Auerbach oral history, CBI OH 2, 5–7.

31. The first UNIVAC to be delivered was finally moved to the Census Bureau in December 1952. The first UNIVAC to be physically moved was serial number two, for the U.S. Air

Force, sent to the Pentagon in June 1952. UNIVAC installations are listed in Ceruzzi, *A History of Modern Computing*, 28.

32. UNIVAC Conference oral history on May 17–18, 1990, Washington, D.C., Charles Babbage Institute OH 200, 76.

33. Snyder, "Atlas and the Early Days of Computers," at tinyurl.com/3btwx9a (March 2011). Snyder assembled forty-five documents on Atlas and other state-of-the-art computing projects; see tinyurl.com/bm848l5 (April 2013).

34. Arnold A. Cohen oral history, CBI OH 58, 33–34.

35. UNIVAC had a basic multiply time of 1,800 microseconds while the ERA 1101, a *slower* commercial version of Atlas (the clock rate was reduced from 1 MHz to 0.4 MHz), could multiply in 260 microseconds. Add times were 120 microseconds (UNIVAC) and 5 microseconds (ERA 1101); all times "excluding storage access." See Weik, *A Survey of Domestic Electronic Digital Computing Systems*, 177, 185.

36. NSA took delivery of two Atlas I models (installed December 1950 and March 1953) as well as two Atlas II models (October 1953 and December 1954), according to Snyder, *History of NSA General-Purpose Electronic Digital Computers*, 93; at tinyurl.com/265zvjn (June 2011).

37. Snyder, "Atlas and the Early Days of Computers," at tinyurl.com/3btwx9a (March 2011).

38. NSA, "Before Supercomputers: NSA and Computer Development" (DOC ID: 3575750), unpaginated, originally "top secret" but declassified in 2003, available at tinyurl.com/cbfdgzc (July 2012).

39. James T. Pendergrass oral history, CBI OH 93, 21.

40. Frank C. Mullaney oral history, CBI OH 110, 67.

41. Walter L. Anderson oral history, CBI OH 119, 37.

42. Marvin L. Stein oral history, CBI OH 90, 10–18, 21–23, 35–42.

43. Arnold J. Ryden oral history, CBI OH 314, 5 (rush to bank); John E. Parker oral history, CBI OH 99, 50. A confidential financial analysis 21 April 1951 in CBI (100), box 1, folder 3, outlines "the need for financing."

44. John E. Parker oral history, CBI OH 99, 57.

45. Arnold A. Cohen oral history, CBI OH 58, 47 (Honeywell), 99 (massive patent specifications); John E. Parker oral history, CBI OH 99, 40–42, 52–56, 82–83.

46. Andrew Russell (Drew) Pearson collection, American University Archives and Special Collections, box 9 folder 20, available at tinyurl.com/3l849e3 (March 2011); Anderson, *Confessions of a Muckraker*, 141; H. Dick Clover oral history, CBI OH 113, 28, 31; John Parker to Francis J. McNamara, January 12, 1952, in CBI (100), box 1, folder 6. "Meader was a difficult man to handle. He didn't get along with Norris. Engstrom was afraid of him. He was, I must say, a trial for me, too. And I finally bought him out," stated John E. Parker (CBI OH 99, 25).

47. Gray, "Engineering Research Associates and the Atlas Computer"; John E. Parker oral history, CBI OH 99, 30. Later, ERA was quietly thanked; see John Parker to George C. Marshall, February 8, 1951, in CBI (100), box 1, folder 10. Besides the muckraking, a rival group within NSA or the navy may have wished to take ERA down a peg. "Almost the only business [ERA] had was with the Navy. And Wenger was satisfied with that arrangement, but other people in the Navy looked at it askance because you know, it looked like Wenger

was just feeding his favorites and they didn't like it. So somebody came along and said it's time that ERA found some other business," stated Howard Campaigne (39–40) in his secret June 28, 1983, NSA interview, now declassified, available at tinyurl.com/6d7dqgg (June 2011).

48. Willis K. Drake oral history, CBI OH 46, 18 (Parker the capitalist); Robert Emmett McDonald oral history, CBI OH 45, 27 (rather inept). The negotiations between Parker and executives at Remington Rand can be followed in CBI (100), box 1, folder 4, and John E. Parker oral history, CBI OH 99, 50–62.

49. Norberg, *Computers and Commerce,* 159–166, 161.

50. Willis K. Drake oral history, CBI OH 46, 16.

51. The narrowly averted "absolute disaster" with the UNIVAC installation at GE–Louisville is related in ibid., 23-31.

52. Robert Emmett McDonald oral history, CBI OH 57, 7.

53. Frank C. Mullaney oral history, CBI OH 110, 60. The UNIVAC and 1101 machines are contrasted in W. G. Welchman to H. T. Engstrom, October 17, 1952, CBI (176), box 6, folder 13.

54. For the Philadelphia–St. Paul tensions, see Arnold A. Cohen oral history, CBI OH 58, 65, 83; Henry S. Forrest oral history, CBI OH 289, 11 (on Groves). See also Robert Emmett McDonald oral history, CBI OH 45, 22–23; George Gray, "The UNIVAC 1102, 1103, and 1104," *Unisys History Newsletter* 6:1 (January 2002), at tinyurl.com/3nkxfje (March 2011); Norberg, *Computers and Commerce,* 217–23; A. R. Rumbles to John Parker, December 12, 1951, in CBI (100), box 1, folder 6; Willis K. Drake oral history, CBI OH 46, 30–35.

55. Norberg, *Computers and Commerce,* 221.

56. Robert Emmett McDonald oral history, CBI OH 45, 19, 24.

57. Frank C. Mullaney oral history, CBI OH 110, 73.

58. Robert Emmett McDonald oral history, CBI OH 45, 32.

59. Marvin L. Stein oral history, CBI OH 90, 10 (intrinsic interest), 11 (Friday afternoon). ERA veteran Erwin Tomash in 1962 organized Dataproducts Corporation, a manufacturer of computer peripherals, and later founded the Charles Babbage Institute at the University of Minnesota. Other Convair computing specialists at the time included Robert M. Price, later CEO of Control Data, and his boss Ben Ferber, later also a CDC executive; see Misa, ed., *Building the Control Data Legacy.*

60. Documentation in the folders "1103 Grant Time and Correspondence 1955–58," box 3, folder 17; "Operating Statistics on the Univac 1103 1958–60," box 1, folder 10; "I.T. Computing Committee 1956–59," box 1, folder 35, all in Marvin Stein papers; Misa and Seidel, eds., *College of Science and Engineering,* 84–95.

61. Marvin L. Stein oral history, CBI OH 90, 17 (extraordinary amount of interest); "1103 Sales Contract," box 2, folder 36, Marvin Stein papers.

3. Corporate Computing

1. Groves, *Now It Can Be Told.*

2. Arnold A. Cohen, CBI OH 58, 64 (lunch-table inventing), 65 (war anecdote); William C. Norris, CBI OH 116, 54 (Thursday lunch with MacArthur). See also Norris, *Racing for the Bomb,* 504–10, 679n17; Bernstein, "Reconsidering the 'Atomic General,'" 883–920; James W. Cortada, *Before the Computer,* 154–56, 233–35.

3. Robert Emmett McDonald, CBI OH 45, 36 (MacArthur image); John Parker, CBI OH 99, 78 (magnificent). MacArthur's military contacts were significant: "General MacArthur would be talking to the commanding general of, say, Rome Air Development Center, so I heard," recalled Walter L. Anderson, oral history, CBI OH 119, 28.

4. William C. Norris oral history, CBI OH 116, 52. On corporate infighting initiated by B. F. Anderson, who was "a good alley fighter . . . in charge of manufacturing," see 52, 74, 76, 79, 80. Persisting problems with engineering–manufacturing cooperation are discussed in W. J. Suchors to R. D. Baker, August 22, 1960, in CBI (176), box 37, folder 6.

5. Arnold A. Cohen oral history, CBI OH 58, 96.

6. Robert Emmett McDonald oral history, CBI OH 57, 8.

7. Neil R. Lincoln et al. on computer architecture and design at CDC, CBI OH 321, 20 (LARC was going so far in the hole); Arnold A. Cohen oral history, CBI OH 58, 83 (Hill leaving).

8. Arnold A. Cohen oral history, CBI OH 58, 86. See also Weik, *A Third Survey of Domestic Electronic Digital Computing Systems*, 958–61.

9. Data from Weik, *A Third Survey of Domestic Electronic Digital Computing Systems*, 992–1001.

10. Lundstrom, *A Few Good Men from Univac*, 22 (abyss), 38 (invasion). On Univac II testing, see "General Information for Conduct of Final Test Univac II," February 14, 1957; D. E. Lundstrom (project engineer for test) memos to test crew, February 14, 1957, March 23, 1957, April 27, 1957, November 8, 1957, all in CBI (176), box 40, folder 3.

11. William C. Norris oral history, CBI OH 116, 57.

12. Arnold A. Cohen oral history, CBI OH 58, 63, 78, 80.

13. Frank C. Mullaney oral history, CBI OH 110, 44.

14. Arnold A. Cohen oral history, CBI OH 58, 80–81.

15. Tomash and Cohen, "The Birth of an ERA."

16. William C. Norris oral history, CBI OH 116, 75. The figure of one hundred per month is from "Antenna Coupler Project in Full Swing at Plant 3," news clipping in Univac scrapbook CBI.

17. Marc Shoquist, "The Antenna Coupler Program," at vipclubmn.org/couplers.aspx (April 2011).

18. Robert Emmett McDonald oral history, CBI OH 45, 17.

19. "Historic Shovel to Break Ground," *St. Paul Pioneer Press*, November 8, 1955, in CBI (80): WCN scrapbook.

20. Carl Hennemann, "Big Site Here Asked by Remington Rand," *St. Paul Dispatch*, April 5, 1955, 1–2; "Ground Rites Set for New Univac Plant" and "St. Paul May Get Electronics Plant" in CBI files.

21. See news clippings "Feminine Invasion of Assembly Shops Successful" [n.d.]; "Antenna Coupler Project in Full Swing at Plant 3"; and news photo "Antenna couplers are assembled at the University Avenue plant" in Leacock collection of ERA and Univac scrapbooks CBI (57). See also Misa, ed., *Gender Codes*, chapter 2.

22. MITRE's photos of Whirlwind are at tinyurl.com/3mfktjz and SAGE at tinyurl.com/94ehum (April 2011). RCA's BIZMAC, a one-of-a-kind computer for the army, at 28,759 tubes is likely the second-largest computer ever built; see Weik, *A Survey of Domestic Electronic Digital Computing Systems*, 159, 229.

23. SAGE computer statistics from Weik, *A Third Survey of Domestic Electronic Digital Computing Systems.* See also R. R. Everett, C. A. Zraket, and H. D. Bennington, "SAGE: A Data Processing System for Air Defense" (brochure written by MITRE for visitors), at ed-thelen.org/sage.html (April 2011). An internal IBM memo, marked up by hand, identified eighty-five thousand instructions per second as the "maximum" instruction rate for the AN/FSQ-7 computer (the transistorized 7A model was somewhat faster); see IBM, "The AN/FSQ 7A Computer" (January 20, 1959) at tinyurl.com/3e7s8ek (April 2011).

24. IBM, "SAGE: The First National Air Defense Network," at tinyurl.com/3k8pe9f (April 2011).

25. For this description, see MITRE, at tinyurl.com/94ehum (April 2011). A 1960 IBM advertisement on YouTube shows the SAGE "situation room" in action: tinyurl.com/bvkqmej (April 2013).

26. Mark Thompson, "A U.S. Defense System Needs East Bloc Parts," *Philadelphia Inquirer* (October 1, 1988), at tinyurl.com/3m4yrt8 (April 2011).

27. Redmond and Smith, *Project Whirlwind;* Baum, *The System Builders;* special issue on SAGE, *Annals of the History of Computing* 5:4 (1981); Jacobs, *The SAGE Air Defense Systems;* Redmond and Smith, *From Whirlwind to MITRE;* Green, *Bright Boys.*

28. See correspondence in the NTDS files in CBI (176) and stories in David L. Boslaugh, unpublished oral history memoir available from Naval Historical Foundation (2002), 29–42.

29. The following paragraphs on NTDS are based on Boslaugh, *When Computers Went to Sea,* 117–265; Paolucci, Polmar, and Patrick, *A Guide to U.S. Navy Command, Control, and Communications;* and Graf, *Case Study of the Development of the Naval Tactical Data System.* Boslaugh has the most extensive treatment of NTDS history.

30. Weik, *A Third Survey of Domestic Electronic Digital Computing Systems,* 61.

31. "UNIVAC-NTDS; UNIVAC 1206; AN/USQ-20," at tinyurl.com/cwfvkmb (April 2013). Thelen credits Cray with the NTDS production computer CP-642. Gordon Bell also specifically credits Cray with the production CP-642 at tinyurl.com/3awnmgk (April 2011). Paul Ceruzzi connects the Cray-designed prototype with the redesigned CP-642 production computer (assuming that the circuit boards of the redesigned production computer, which he examined in Boston and is now in Mountain View, are Cray's design); see Ceruzzi, "'The Mind's Eye' and the Computers of Seymour Cray," 151–60. The Computer History Museum also credits a 1962-vintage NTDS production computer to Cray at tinyurl.com/44eqay3 (June 2011). The designer of the NTDS printed circuit cards was Lee Granberg, according to Boslaugh, *When Computers Went to Sea,* 219. Cray's comment about the 1103 is from his Smithsonian interview with David Alison; see tinyurl.com/bpbowzm (April 2013).

32. Boslaugh, *When Computers Went to Sea,* 360.

33. The Control Data 6600 (1964) with its seventy-four-item instruction set led some observers to proclaim that RISC stands for "Really Invented by Seymour Cray." By comparison, the IBM 7030 Stretch had 154 basic instructions.

34. Weik, *A Third Survey of Domestic Electronic Digital Computing Systems,* 43.

35. A later transistorized version, AN/FSQ 7A, did notably feature "inter-machine communication between memory blocks" but the vacuum-tube version deployed in the field for SAGE did not. See IBM, "The AN/FSQ 7A Computer" (January 20, 1959), at tinyurl.com/3e7s8ek (April 2011).

36. WCN to H. S. Forrest, September 6, 1957, September 9, 1957, October 16, 1957, December 6, 1957; in CBI (80), series 9, box 28, folder 1: WCN day files.

37. The unit computer itself was labeled as CP-642 while the computer system with peripherals was AN/USQ-20. Univac sold a military version of the NTDS computer as the Univac 490 and a commercial version as the Univac 1206. For details of NTDS testing, relations with the NEL (San Diego), and shipboard tests, see Jay Kershaw papers CBI (176), box 40, folder 1.

38. Boslaugh, *When Computers Went to Sea*, 221.

39. Remington Rand Univac (St. Paul Division), "Naval Tactical Data System Technical Note No. 244: AN/USQ-20 Unit Computer Characteristics" (October 10, 1960), at tinyurl.com/3rx9pxu (April 2011).

40. Manufacturing and delivery schedules can be followed in Vern E. Leas to Capt. Ed Svendsen, January 4, 1961, and February 21, 1961; all in CBI (176), box 37, folder 6.

41. News clippings, "The Story Is Growth" (April 1967); "Construction Starts on New Univac DSD Facility in Eagan Township" (August–September 1966); "Ground Broken for New DSD Office, Lab Near St. Paul," n.d.; "Year-End Report Shows [Univac] Employment at 10,500," [1968]; all in Leacock collection of ERA and Univac scrapbooks CBI (57).

42. Lundstrom, *A Few Good Men from Univac*, 62.

43. Remington Rand Univac, "Naval Tactical Data System Technical Note No. 244: AN/USQ-20 Unit Computer Characteristics," 11.

44. Ralph A. Hileman (Military Department, San Diego Engineering Center), "The Naval Tactical Data System Modular Concept," paper presented to Institute of the Aerospace Sciences, January 22–24, 1962, in CBI (176), box 40, folder 6. Two Univac papers were selected for the 1961 Eastern Joint Computer Conference—according to G. G. Chapin memo, July 25, 1961, in CBI (176), box 37, folder 6—and belatedly published as Chapin, "Organizing and Programming a Shipboard Real-Time Computer System," 127–37; and Pickering, Mutschler, and Erickson, "Multicomputer Programming for a Large Scale Real-Time Data Processing System," 445–61.

45. The seven-person NTDS design staff consisted of five St. Paul engineers and two San Diego navy officers, including defined communication responsibilities with each of the seven contributing companies (Hughes, Stromberg-Carlson, Collins Radio, Daystrom/Manson, Bell Telephone Laboratories, Benson-Lehner, and Univac); see G. G. Chapin memo, October 22, 1959, in CBI (176), box 37, folder 1. On the Hazeltine "fix," see C. W. Glewwe record of phone conversations, August 18–21, 1961, with Bert Whitehouse and Bob Aitken (Hazeltine), in CBI (176), box 37, folder 6.

46. By 1980, there would be a total of ninety-three NTDS warships. Eventually, NTDS was incorporated into the Aegis Combat System; NTDS lost its identity as a separate system and became a dependent subsystem. The first Aegis-based warship was the USS *Ticonderoga* (commissioned in 1983).

47. Quotes from Jim Seaman, "UNIVAC-NTDS–Vietnam" (December 2010), at tinyurl.com/c29n3vv (April 2013). See also Boslaugh, *When Computers Went to Sea*, 347–57; Treadway, *Hard Charger*, 71–135.

48. "No computer presently exists . . . [owing to] the problem of obtaining continuous error-free operation of a high-speed computing machine," noted an early study by David R. Israel, "The Application of a High-Speed Digital Computer to the Present-Day Air Traffic Control System," 13.

49. See "Civil Aeronautics Administration Online Input/Output System 1957–1959," in CBI (176), box 39, folder 1; Remington Rand Univac, "Real-Time Computer Systems for Air Traffic Control and Air Line Reservation Applications" (June 1960), in CBI (176), box 40, folder 1. Technically, the replacement computer was the Univac 8303 Input Output Processor.

50. A critical insider's account of the FAA's Advanced Automation System is Britcher, *The Limits of Software*, 149–89.

51. General Accounting Office, *Air Traffic Control*; Stix, "Aging Airways," 96–104; Willemssen, *Year 2000 Computing Crisis*; Government Accountability Office, *National Airspace System*; "Air Traffic Control Systems (ATC)," vipclubmn.org/aircontrol.aspx (May 2011); Brusehaver, "Linux in Air Traffic Control."

52. "A couple of years ago I managed to wangle a tour of the FAA's Air Traffic Control Center at Tulsa, Oklahoma. The controller that showed me around and I had some common military experiences so he was prone to showing me some things that were not on the normal tour. There was a door marked 'Restricted Area–No Entry'. He did not allow me to enter, but he did open the door and let me look in. There were several 'really old' computers, which looked to me like [NTDS-issue] CP-642s with the 'refrigerator' look and the control panel at the top. His comment at the time was that the FAA was still using fifty-year-old computer technology for air traffic control and what a pain it was to find people who were familiar with the programming," states Jim Seaman, "UNIVAC-NTDS-Vietnam" (December 2010), at tinyurl.com/c29n3vv (accessed April 2013).

53. Lundstrom, *A Few Good Men from Univac*, 61.

54. See quotations from NEL Evaluation Report #1123 in CBI (176), box 39, folder 12; Weik, *A Fourth Survey of Domestic Electronic Digital Computing Systems*, 286.

55. Holbrook, "Controlling Contamination," 173.

56. Anderson and Peterson (Remington Rand Univac), "Analysis of Random Failures," 242–53; the journal *Microelectronics and Reliability* (1962–) traces the published scientific literature. See also Anderson, "Failure Modes in High Reliability Components," 359; Selikson and Longo, "A Study of Purple Plague and Its Role in Integrated Circuits," 1638–41; Philofsky, "Purple Plague Revisited," 177–85; Boslaugh, *When Computers Went to Sea*, 221–28; Lojek, *History of Semiconductor Engineering*, 208–13; Holbrook, "Controlling Contamination," 173–91.

57. I am indebted to Mike Svendsen, who was involved with Univac's quality assurance in the 1960s and directed the SCF from 1974 to 1981. He let me examine the detailed internal reports, striking photographs, and vendor presentations that provide an industry-wide assessment of semiconductor quality. "There is only minimal existing/available archival documentation on the history of the SCF," was the earlier assessment of Jeffrey Yost in "Manufacturing Mainframes," 234n60.

4. Innovation Machine

1. See *Electronics* (Business Edition) 30:10B (October 20, 1957): 1, 10. On Sputnik and the founding of ARPA, and its impact on computing, see Norberg and O'Neill, *Transforming Computer Technology*, 5–23.

2. The 1604 story is scrutinized and *rejected* as erroneous "folklore" by a group of CDC computer engineers; see Neil R. Lincoln et al. on computer architecture and design at CDC,

CBI OH 321, at purl.umn.edu/104327, 20–21. Mike Pavlov: "Along the line of Folklore I have heard . . . concerning the naming of the [CDC] 1604 . . . is it true that the name 1604 was derived from 501 Park Avenue being added to Univac 1103? *(Laughter).*" Chuck Hawley: "It was quite popular at the time that this was the origin of 1604." Dolan Toth: "We've never been able to substantiate it. However, there's still lots of people who believe it." Hawley: "The official word promulgated by Mullaney, Cray, and Thornton was that 1604 stood for 16K central memory and 4 tape units. That was the original design goal for the 1604. And then the 1604 went to a 2-bank memory so it ended up with 32K, but it started out as a 16K machine."

3. Murray, *The Supermen,* 66. "I didn't start Control Data. I mean I wasn't part of forming the corporation," according to Seymour Cray oral history, Smithsonian Institution at tinyurl.com/bpbowzm (April 2013). Norris's letters indicate that Cray's hiring was discussed from September 1957 on.

4. Norberg, *Computers and Commerce,* 263, 273.

5. Willis K. Drake oral history, CBI OH 46, 36.

6. Frank C. Mullaney oral history, CBI OH 110, 75 (boonies); Willis K. Drake oral history, CBI OH 46, 36 (sledgehammers).

7. Both the Ryden and Drake interviews contradict James Worthy's claim that "Norris . . . made all the substantive decisions himself. An early one was naming the new company. . . . *Control Data Corporation* was the [name] Norris liked best, and that was the one selected" (Worthy, *William C. Norris,* 32).

8. Ryden oral history, CBI OH 314, 10.

9. Ibid., 22; WCN to WCN *[sic],* August 13, 1957, CBI (80), series 9, box 35: WCN scrapbook; Sterling A. Barrett to WCN, July 18, 1957, CBI (80), series 9, box 8, folder 15.

10. Willis K. Drake oral history, CBI OH 46, 47; articles of incorporation are in CBI (80), series 8, box 24, folder 3. Also in folder 3 there is a separate "proposal to W. C. Norris" outlining his position as president, with salary, stock options, and benefits. Norris resigned from Univac on July 26, according to T. C. Fry to F. C. Mullaney, July 26, 1957, CBI (80), series 9, box 35: WCN scrapbook. Sperry Rand threatened legal action in Norman B. Frost to Control Data Corporation, August 6, 1957, CBI (80), series 9, box 28, folder 1, WCN day files.

11. Seymour Cray oral history, Smithsonian Institution, 13, at tinyurl.com/bpbowzm; Raymond W. Allard oral history, Charles Babbage Institute OH 288, [n.p.]; Lyle K. Anderson *(Minneapolis Star and Tribune)* to Control Data Corporation, August 7, 1958, CBI (80), series 8, box 24, folder 4: Lease–501 Park Avenue.

12. Willis K. Drake oral history, CBI OH 46, 48, 54; CDC–UMDC agreement canceling lease, August 7, 1958, CBI (80), series 8, box 24, folder 4: Lease–501 Park Avenue. The July prospectus is quoted in Price, *The Eye For Innovation,* 4.

13. Ryden oral history, CBI 314, 12; Thomas G. Kamp oral history, CBI OH 297 [n.p.] (not online); Eugene Baker oral history, CBI OH 293 [n.p] (not online). Many dates can be verified in the fourteen boxes of Control Data Corporation Records: News Releases, 1957–90, CBI 80 series 13. WCN sought business from his contacts at NSA (Joe Eachus and Howard Engstrom, then deputy director) and other companies and federal agencies.

14. Paul J. Bulver oral history, CBI OH 279 [n.p.]; Paul J. Bulver oral history, CBI OH *300* [n.p.] (cash cow, springboard, Swiss technicians); Lloyd M. Thorndyke oral history, CBI OH 280 (technical work at Cedar).

15. Leo F. Slattery oral history, CBI OH 284 [n.p.] (chaotic); James Harris oral history, CBI OH 295, 1 (knuckles); Frank C. Mullaney oral history, CBI OH 110, 82 (drill).

16. See "Major Minnesota Missile Industry Goal of New Group," *Electronic News* (March 3, 1958); "Control Data Employment Up," *St. Paul Dispatch* (March 14, 1958); "Company Gains Defense Work," *Minneapolis Star* (March 14, 1958); WCN to Henry Forrest, March 23, 1958; R. H. Thuleen to A. J. Ryden, April 9, 1958; "Control Data Retains 'U' Consultant," *Minneapolis Star* (September 23, 1958), 15A; all in CBI (80), series 9, box 35: WCN scrapbook. CBI (80), series 8, box 24, folder 13: W. G. Shepherd–Consultant Agreement (1958); WCN to W. G. Shepherd, March 13, 1958, April 28, 1958; in CBI (80), series 9, box 28, folder 1: WCN day files. Shepherd was paid $150 a month in exchange for twelve days per calendar year and duties as associate director of research of CDC; he was offered an option for one thousand shares of CDC stock.

17. Frank C. Mullaney oral history, CBI OH 110, 83. By December 1957, the U.S. Naval Postgraduate School, which soon bought the first CDC 1604, was "requesting information about Seymour Cray's computer"; see WCN to H. S. Forrest, December 6, 1957, in CBI (80), series 9, box 28, folder 1: WCN daybooks.

18. Ryden oral history, CBI 314, 14–20, 24–25; Willis K. Drake oral history, CBI OH 46, 41–44.

19. Hsu and Kenney, "Organizing Venture Capital," 579–616.

20. Ryden oral history, CBI 314, 14–20, 24–25; Willis K. Drake oral history, CBI OH 46, 46, 56–60; "Firm Set Up to Help Technical Industries," *Minneapolis Star* (October 23, 1958); Willis K. Drake, "A Look at the Local Scene," *Upper Midwest Investor* 1:1 (April 1961): 20.

21. As Ryden oral history, CBI 314, recalled, "We were organizing Midwest Tech, but the regulations weren't out yet on the SBICs [the Small Business Investment Act of 1958 permitted Small Business Investment Companies] so we went ahead under the Investment Company Act of 1940 even though there were reasons why the SBICs were more attractive. . . . They sued all of the officers and directors, about 20 people. To shorten the story, we were exonerated after a five week trial in January and February of 1963, but it wasn't until late in that year that the judge gave his verdict. He threw the SEC a bone in that none of us could be officers or directors of funds under the investment company act without prior SEC approval. So we all resigned [from MTDC]" (17, 25). On Telex, see Wirtzfeld, "Telex, Inc.," 6–8, 21. On MTDC insiders, see *Tomash v. Midwest Technical Development Corp.* 160 NW 2d 273 Minnesota Supreme Court (1968).

22. See Henry S. Forrest oral history, CBI OH 289, 9.

23. CDC prospectus in CBI (80), series 8, box 24, folder 3; CDC contract awards are identified CBI (80), series 14: newspaper and magazine articles, 1955–87, at purl.umn.edu/41224; Lawrence Livermore National Laboratory, "Stories of the Development of Large Scale Scientific Computing: Page 4: Significant Hardware and Software Work," at tinyurl.com/7cvvv5f (January 2012).

24. Leo F. Slattery oral history, CBI OH 284 [n.p.].

25. Elmer Andersen to WCN, March 1, 1961; "Computer Firm Picks New Site," *St. Paul Pioneer Press* (February 25, 1961); "Control Data Settles on Site in Bloomington," *Minneapolis Star* (February 25, 1961); "Roseville + Bloomington + Control Data = Problem for Suburbs Seeking Industry," *Minneapolis Star* (March 2, 1961); all in CBI (80), series 9, box 35: WCN

scrapbook; Mozingo, *Pastoral Capitalism;* CDC, "New Plant Site: A Survey and Recommendations" (August 1, 1960), in CBI (80), series 8, box 24, folder 6: New Building Roseville [sic] 1960.

26. "15-Million-Dollar Project Planned for Bloomington–Edina Area," *Minneapolis Star* (March 1961); "Control Data Predicts Still Further Growth," *Minneapolis Morning Tribune* (September 20, 1961); in CBI (80), series 9, box 35: WCN scrapbook.

27. Marketing chief "George Hanson raised strong objections to CDC's [OEM] selling of its high-performance peripherals to its competitors," according to Lloyd M. Thorndyke oral history, CBI OH 303 [n.p.]. The peripheral products business was also given grief by the computer systems group. "There was a lot of friction between the peripherals business and the computer business. . . . Cedar [was] sending its paper tape readers for a quality check at 501 Park [downtown]. They were always rejected because the view glass was always falling out. It was discovered that the QA [quality assurance] manager had been leaning on the glass with his thumb as hard as he could to pop them out," recalled Eugene Baker oral history, CBI OH 293 [n.p.].

28. Thomas G. Kamp oral history, CBI OH 297 [n.p.]; CBI 80 series 13, subject listings for "peripherals" 1960–66; Control Data Corporation Records: Research and Development Project Reports, 1959–64, CBI 80, series 25, box 14, folder 145: Cedar Engineering, Project 5066, July 15, 1959 to June 30, 1961.

29. Gordon R. Brown oral history, CBI OH 294 [n.p.]; Takahashi, "The Rise and Fall of Plug-Compatible Mainframes," 4–16. In his *A History of Modern Computing,* Ceruzzi focuses on the IBM System/360 in his treatment of "plug compatibles" (161–65) but clearly the IBM 1401 was a target, too. Lloyd Thorndyke identifies the birth of "plug compatibles" with an early Control Data design to gain an OEM sale to NCR substituting an Ampex tape unit (Lloyd M. Thorndyke oral history, CBI OH 303 [n.p.]).

30. Thomas G. Kamp oral history, CBI OH 297 [n.p.]; Ceruzzi, *A History of Modern Computing,* 124–41, 191–93; Lloyd M. Thorndyke oral history, CBI OH 303 [n.p.].

31. "Each 1604 was built as a project—a separate room, and there was a project engineer and several technicians assigned to that particular machine, but there were some people who did a kind of manufacturing process in that they would do the backboard work elsewhere and the panels would be brought in. The final assembly of the things in the machine was done by this group of people, one of whom was the customer engineer who was going to go out with that system to the field," stated Richard C. Gunderson oral history, CBI OH 266, 22.

32. Thomas G. Kamp oral history, CBI OH 297 [n.p.]; Weik, *A Fourth Survey of Domestic Electronic Digital Computing Systems,* 80–81; Leo F. Slattery oral history, CBI OH 284 [n.p.] (physical and logical size of 16 KB memory). Because the computer's word length was typically forty-eight or sixty bits, the total number of binary digits was significantly larger than sixteen KB with today's standard eight-bit "words."

33. Richard C. Gunderson oral history, Charles Babbage Institute OH 266, 1; James D. Harris oral history, Charles Babbage Institute OH 295; Charles F. Crichton oral history, Charles Babbage Institute OH 264 [n.p.] (not online).

34. "At that time, the hotbed of programming talent was in California—the west coast in general, and California in particular. . . . It was all because of the aerospace industry," stated Robert Price (Misa, ed., *Building the Control Data Legacy,* 37).

35. Richard A. Zemlin oral history, Charles Babbage Institute OH 152, 4; Richard A. Zemlin oral history, Charles Babbage Institute OH 307 (not online); Misa, *Building the Control Data Legacy*, 56–59.

36. Richard Zemlin oral history, Charles Babbage Institute OH 152, 6 (Co-op Monitor), 21 (condition of the contract); Richard Zemlin oral history, Charles Babbage Institute OH 307 [n.p.] (describes Co-op Monitor) (not online).

37. Robert M. Price, oral history with the author, in Misa, *Building the Control Data Legacy*, 46; Mike Schumacher oral history, Charles Babbage Institute OH 287, 2–3 (not online); Raymond W. Allard oral history, Charles Babbage Institute OH 288, [n.p.] (not online).

38. James Harris oral history, CBI OH 295, 4 (sacred cows; accounting), 18 (corporate orphans); Charles Crichton oral history, CBI OH 264 [n.p.]; Paul J. Bulver oral history, CBI OH 300 [n.p.] (selling Cedar). Harris on accounting: "There were lots of ways of making money in those days because we used some real Mickey Mouse accounting where you get credit for internal sales and if you sold anything why, you know, you'd mark it up so far it was astronomical. All these things went on, and we had some . . . well, we called it 'funny money' . . . in our income revenue records." In 1965, "every single division in Control Data made money [but] by the time they eliminated interdivisional profit, the corporation lost $3 a share" owing to such accounting tricks as "phase-of-the-moon artificial profit," according to Lloyd M. Thorndyke oral history, CBI OH 303 [n.p.].

39. Herbert W. Robinson oral history, Charles Babbage Institute OH 147.

40. See Control Data Corporation Records (CBI 80, Series 1) Acquisitions, Subsidiaries, and Joint Ventures, 1951–91, at purl.umn.edu/41253 (June 2011); and Price, *The Eye for Innovation*, 158–75.

41. Charles F. Crichton oral history, Charles Babbage Institute OH 264 [n.p.] (not online).

42. See Martin Dodge's "Historical Maps of Computer Networks," at tinyurl.com/yjykjyh (June 2010). In September 1971, ARPANET had eighteen centers each with a minicomputer creating a full-fledged node; in addition, there were eight terminals with access to the network.

43. "Tackling IBM," *Time* (December 20, 1968), at tinyurl.com/63lusvq; "A Settlement for IBM," *Time* (January 29, 1973), at tinyurl.com/y8q32w9 (June 2011).

44. On the Control Data–IBM lawsuit database, see Elmer B. Trousdale oral history, CBI OH 271, 28–34, 45–47, 30 (damaging documents); Richard G. Lareau oral history, CBI OH 273; Robert M. Price, oral history with author, July 1, 2009, July 8, 2009.

45. This Cray plant anecdote from Thomas G. Kamp oral history, CBI OH 297 [n.p.]. There were other, less dramatic episodes of autonomous design groups: "In Control Data, there was for a long period of time compartmentalization in design labs. Small groups of people would go off by themselves and do their thing. The best example of that is the group that eventually went to Chippewa. There are other examples of that too," according to Leo F. Slattery oral history, CBI OH 284 [n.p.].

46. Frank C. Mullaney oral history, CBI OH 110, 67.

47. Leo F. Slattery oral history, CBI OH 284 [n.p.].

48. In his foreward to Thornton's book (*Design of a Computer*, i), Cray wrote: "The reader can rest assured that the material presented is accurate and from the best authority as Mr. Thornton was personally responsible for most of the detailed design of the Control Data model 6600 system." Cray quotes in this and the next paragraph from Seymour Cray oral history, Smithsonian Institution, 13, 23.

49. Dimitry Grabbe oral history, IEEE Global History network, at tinyurl.com/3c3nkjy (July 2011).

50. Watson, *Father, Son and Co.*, 383; see original at tinyurl.com/3h25weo (June 2011).

51. For instance, while most public sources suggest sale of just one CDC 1604 to the NSA, the NSA's own (now declassified) history of computing developments through 1963 indicates that the NSA purchased five CDC 1604s, several with special cryptographic hardware additions (Snyder, *History of NSA General-Purpose Electronic Digital Computers*, 78–82).

52. The 8.7 hour figure was for the Univac 1 installed at the Department of the Air Force (the same machine used for the 1952 election), according to Weik, *A Survey of Domestic Electronic Digital Computing Systems*, 177–81. Computing speeds are from Lawrence Livermore National Laboratory, "Stories of the Development of Large Scale Scientific Computing," (n.d.), at tinyurl.com/7cvvv5f (April 2013).

53. Reilly, *Milestones in Computer Science and Information Technology*, 41 (0th to NSA); Jitze Couperus (Control Data Systems), at tinyurl.com/6jp9ug8 (July 2011) (anonymous truck).

54. Compare Buchholz, ed., *Planning a Computer System*, 254–71, with Snyder, "Influence of U.S. Cryptologic Organizations on the Digital Computer Industry," 79, at tinyurl.com/3q9fc6a (July 2011).

55. On the cryptographic instructions built into ERA's Atlas computers, see Snyder, *History of NSA General-Purpose Electronic Digital Computers*, 10, 13, 96. The "vector add" instruction was essentially a binary "add" *without* carry that was useful for quickly comparing strings of data; it is not the vector addition used in engineering and physics.

56. NSA, "Before Supercomputers: NSA and Computer Development" (DOC ID: 3575750), originally "top secret" but declassified 2003, available at tinyurl.com/cbfdgzc (July 2012).

57. See the discussion about the 1604's "77" instruction, as a "perform algorithm" instruction, in CBI OH 321, 32. Two of the NSA's five CDC 1604s were "equipped with an additional powerful instruction, the '77' Order. By selecting suitable patterns of ones and zeros, tremendous numbers of variations of transmissive and arithmetic operations can be called into play, involving CDC-1604 arithmetic registers and index registers. This instruction increases their flexibility for analytic applications" (Snyder, *History of NSA General-Purpose Electronic Digital Computers*, 82). For commercial 1604s, instruction 77 was "not used." At least two other NSA 1604s were equipped with special hardware additions, WELCHER and PULLMAN.

58. See Ed Thelen's entry on the CDC 6600 at tinyurl.com/3rzlk44. On the "weight'" of binary numbers, see Wegner, "A Technique for Counting Ones in a Binary Computer," 322. For the 6600's "population count," see central-processor instruction 47 "sum of 1's in Xk to Xi [up to eight 60-bit operand registers]" to be completed in 8 100-nanosecond minor clock cycles (in Thornton, *Design of a Computer*, book front end sheets, 14, 16). Don Pagelkopf indicated that CDC 1604 serial 30 was sent to NSA with "sideways add" or "pop count" and "bit merge" instructions in Control Data engineers oral history, CBI OH 321, 32.

59. See Hamming, *Coding and Information Theory*, 44–50.

60. "Control Data and the North Side," *Minneapolis Tribune* (November 28, 1967); Dick Cunningham, "Control Data Plans Plant on North Side," *Minneapolis Tribune* (November 28, 1967); "Industry Moves to the Ghetto," *Minneapolis Star* (November 28, 1967).

61. William C. Norris CBI OH (299), [n.p.], September 23, 1977 interview.

62. Sterling A. Barrett to WCN, July 18, 1957, CBI (80), series 9, box 8, folder 15; WCN to Sterling A. Barrett, October 18, 1967 (funnel), January 9, 1968, February 7, 1968 (ghetto problem); WCN to N. R. Berg, December 8, 1967; CBI (80), series 9, box 29, folder 12, WCN day files; Jack Wilson, "U.S. Awards Control Data $1 Million to Train Jobless," *Minneapolis Tribune* (February 24, 1968), 1.

63. Hart, "From 'Ward of the State' to 'Revolutionary without a Movement.'" 217; WCN to H. P. Donaghue, April 12, 1968 (no. 1 problem); WCN to CDC executives, May 10, 1968 (Ginzberg meeting); WCN to Clemmer G. Wait, May 21, 1968 (listen to his views); in CBI (80), series 9, box 29, folder 13: WCN day files.

64. WCN to Donald M. Graham, April 12, 1968; WCN to Harold LeVander, April 15, 1968; WCN to H. D. Barnard, May 13, 1968; in CBI (80), series 9, box 29, folder 13: WCN day files.

65. Employment numbers from Frank Dawe, May 6, 1990, in CBI (80), series 9, box 30, folder 18: WCN day files. On the Microelectronics and Computer Technology Corporation, see Gibson and Rogers, *R&D Collaboration on Trial*.

66. Norris once gave a speech estimating the *total* sum of money invested by NSF, the University of Illinois, Control Data, and others in the hundreds of millions. This figure, erroneously but ubiquitously, became the sharp critical judgment: "He spent a billion dollars on a computer-based education system"; see Walter Kirn, "William C. Norris: The Bleeding-Heart Rationalist," *New York Times* (December 31, 2006), at tinyurl.com/7u9egld.

67. Jonathan Weber, "Silicon Graphics Will Unveil Desk-Side Computer with the Power of a Cray," *Los Angeles Times* (January 27, 1993), available at tinyurl.com/88frd8z (July 2012).

5. First Computer

1. Ross Jones, "Malcolm Moos and the Military–Industrial Complex Speech," June 14, 2011, Johns Hopkins e-Libris, at tinyurl.com/7tst5se (January 2012); Jim Newton, "Ike's Speech," *New Yorker* (December 20, 2010), at tinyurl.com/3ytbwya (January 2012).

2. Nessell, *Honeywell*, 8–11.

3. Rodengen, *The Legend of Honeywell*, 35–36, 4.

4. Nessell, *The Restless Spirit*, 108. The working of Norden bombsights is made clear in Volta Torrey, "How the Norden Bombsight Does Its Job," *Popular Science* (June 1945): 70–73. Oddly, some accounts confuse the Norden *bombsight* with the ENIAC's calculations for *artillery shells*; see Mark Wolverton, "Girl Computers," *American Heritage* 612 (2011), at tinyurl.com/7uyzqs6 (July 2012); and LeAnn Erickson, "Top Secret Rosies: The Female Computers of WWII" (2010), at www.topsecretrosies.com.

5. See this suggestion for Elmer Sperry in Hughes, *Elmer Sperry*.

6. Rodengen, *The Legend of Honeywell*, 92–101 (civilian and military space); Arnold A. Cohen oral history, CBI OH 58, 53 (Boeing 707).

7. Smith, "A New Large-Scale Data-Handling System, DATAmatic 1000," 22–28.

8. Honeywell ad in *Datamation* 65 (September 1965): 48; Strassmann, *The Computers Nobody Wanted*, 6. For IBM Captivator, see tinyurl.com/4xz07zq (June 2011).

9. Louis Fein oral history, CBI OH 15.

10. See Adrian Wise, "Computer Control Company," at tinyurl.com/44724j5 (August 2011).

11. Atkinson, "The Curious Case of the Kitchen Computer," 173.

12. John W. Mauchly, deposition, October 11–13, 1967, in CBI (1), box 47, folder 6, 44.

13. Preliminary work on the ENIAC patent began as early as September 1944, according to the detailed first-person account of McTiernan, "The ENIAC Patent," 54–58, 80. The total of eleven patent interferences filed against the ENIAC patent, as well as the subsequent legal proceedings, can be followed in Judge Earl Larson, "Findings of Fact, Conclusions of Law, and Order for Judgement," CBI (1), box 51: folder 2, 76–85 (finding 11 on Delay).

14. Reich, *The Making of American Industrial Research,* 138.

15. Irvine, "Early Digital Computers at Bell Telephone Laboratories," 22–42. There have been sightings of a (never-published) computing volume still in the Bell Labs archives. In addition to the Stibitz and other computing patents, Bell Labs achieved a Nobel Prize for inventing the transistor (1947) as well as built an early transistorized computer (1954), modem (1960), the operating system UNIX (1969), and the programming languages C (1973) and C++ (1983).

16. Ceruzzi, *Reckoners,* 76–93; Stibitz and Loveday, "The Relay Computers at Bell Labs: Part 1," 35–44. See these patents—all filed with assignment to Bell Telephone Laboratories—in George Stibitz, "Complex Computer," U.S. Patent 2,668,661, filed April 19, 1941, issued February 9, 1954; Stibitz, "Counting Device," U.S. Patent 2,292,489, filed November 26, 1941, issued August 11, 1942; Stibitz, "Card Translator," U.S. Patent 2,361,246, filed May 5, 1943, issued October 24, 1944; Stibitz, "Automatic Calculator," U.S. Patent 2,666,579, filed December 26, 1944, issued January 19, 1954; Stibitz, "Biquinary System Calculator," U.S. Patent 2,486,809, filed September 29, 1945, issued November 1, 1949. Stibitz also filed patents independently after leaving Bell. Errors in scanning make it difficult to search for the Complex Computer on Google Patents; see instead www.google.com/patents/US2668661.

17. Ceruzzi, *Reckoners,* 93–101; Stibitz and Loveday, "The Relay Computers at Bell Labs: Part 2," 45–49; Hamming, "Error Detecting and Error Correcting Codes," 147–60; Blackman, *A Survey of Automatic Digital Computers,* 9–10. The interference cases against the ENIAC patent is described in Judge Earl Larson, "Findings of Fact, Conclusions of Law, and Order for Judgement," CBI (1), box 51, folder 2, 76–85 (finding 11 on Delay). The first patent interference was filed in March 1951 by NCR, and the third interference was filed in May 1952 by Bell Telephone Laboratories. Additional interferences were filed by Stromberg-Carlson, IBM, and others. Bell and Stibitz had a falling-out after he left Bell in 1945. Bell instead relied on its Samuel B. Williams patent; and in fact Stibitz later in the 1950s worked half-time for Sperry Rand *against* his former employer (Larson finding 11.8.6).

18. Judge Earl Larson, "Findings of Fact, Conclusions of Law, and Order for Judgement," CBI (1), box 51, folder 2, 76–85 (finding 11 on Delay).

19. See Iowa State University Research Foundation, Inc., Intervenor-appellant, v. Honeywell, Inc., Plaintiff-appellee, v. Sperry Rand Corporation et al., Defendants-appellees 459 F.2d 447 (May 3, 1972). McTiernan, "The ENIAC Patent," 57 (greedy).

20. The legal wrangling can be followed in Illinois Scientific Developments, Inc., Petitioner, v. Honorable John J. Sirica, Judge, United States District Court for the District of Columbia, Respondent, Honeywell, Inc., Intervenor 410 F.2d 237 (January 10, 1968).

21. For these records, see ENIAC Trial Exhibits Master Collection (CBI 145), at purl.umn.edu/40695. Honeywell's perspective can be found in Honeywell vs. Sperry Rand Records, 1925–1973 (CBI 1), at purl.umn.edu/40608, and Charles W. Bradley collection on

the ENIAC Trial (CBI 140), at purl.umn.edu/53937. Sperry Rand's perspective is available at Sperry Rand Corporation, Univac Division, Honeywell vs. Sperry Litigation Records (CBI 72), at purl.umn.edu/41265; and the large collection at Hagley, Sperry–UNIVAC Company Records, at tinyurl.com/7yj6kgj (January 2012). The presiding judge's papers are Earl R. Larson Papers (CBI 31), at purl.umn.edu/40561. Records from Penn's Moore School are ENIAC Patent Trial Collection, 1864–1973, at tinyurl.com/2e86l (January 2012).

22. See Mauchly's deposition, October 11–13, 1967, in CBI (1), box 47, folder 6, 7–9, 19 (lawyers debating relevant questions), 41–42 (NSA product).

23. Earl Larson to Arthur Burks, December 4, 1984, CBI (31), box 1, folder 1.

24. Judge Earl Larson, "Findings of Fact, Conclusions of Law, and Order for Judgement," CBI (1), box 51, folder 2, 50 (finding 4.2.1 on Burks), 54 (findings 4.3.23–28 on "heavy burden" of Honeywell). See Burks and Burks, *The First Electronic Computer*, and Burks, *Who Invented the Computer?*. Larson describes his work on the decision in a friendly letter to John V. Atanasoff, June 14, 1984, CBI (31), box 1, folder 1.

25. Judge Earl Larson, "Findings of Fact, Conclusions of Law, and Order for Judgement," CBI (1), box 51, folder 2, 47 (finding 3.1.2). Garbled and misquoted versions of this finding abound.

26. United States Patent and Trademark Office, 2137 35 U.S.C. 102(f)-2100 Patentability; at tinyurl.com/846njbt (January 2012).

27. John W. Mauchly, deposition, CBI (1), box 47, folder 7, 88–95 (visit to JVA).

28. Judge Earl Larson, "Findings of Fact, Conclusions of Law, and Order for Judgement," CBI (1), box 51, folder 2, 47–49 (finding 3 on Atanasoff), 47 (breadboard, pilot model), 48 (well advanced, free and open), 49 (analog calculating device and digital computer).

29. John W. Mauchly trial testimony, November 9, 1971, CBI (1), box 52, folder 4, 11,807 (saw Stibitz computer at Dartmouth meeting) and 11,847 ("computing circuits" diagrams).

30. For details on the ABC reconstruction, see John Gustafson, "Reconstruction of the Atanasoff-Berry Computer" (1999), at tinyurl.com/8x7fequ (January 2012). For YouTube video, see tinyurl.com/2507apd (April 2013).

31. Marv Davidov oral history, by Clarke A. Chambers, September 27, 1995, 28, 30, at purl.umn.edu/49116 (August 2011). In interviews, Davidov sometimes named 1970 as the year of the spring meeting with CEO Jim Binger. The St. Louis Park Historical Society identifies the spring meeting as occurring in 1969, with the first protests at Honeywell's downtown headquarters in April and leafleting at its St. Louis Park factory in May 1969; for the St. Louis Park Historical Society's materials on the Honeywell Project, see tinyurl.com/7d2938e (January 2012).

32. See tinyurl.com/7hp46vu (January 2012).

33. Marv Davidov oral history, by Clarke A. Chambers, September 27, 1995, 28 (conveyor belt), at purl.umn.edu/49116 (August 2011); Marv Davidov, "Report from the Honeywell Project" (Mss. Minneapolis, June 1969), 4 (funny stuff), copy in author's possession.

34. Dayton quoted in "Pulse of the Twin Cities," by Sid Pranke, February 24, 2006, at tinyurl.com/77kqj9w (January 2012).

35. FBI activities are stressed in Masters and Davidov, *You Can't Do That!;* Lynn Brockman, "Social Movement Repertoires and Dynamics," 136 (Midwest way); the "broken windows and mace" quote is from the St. Louis Park Historical Society's Honeywell Project materials.

36. Jeremy Iggers, "Alliant Is Firm in Not Engaging the Rhetorical Enemy," *Minneapolis Star Tribune* (August 28, 2004), at tinyurl.com/7yz8ve9; Roxana Tiron, "ATK Is Moving Headquarters to Virginia for Proximity to Pentagon," *Bloomberg News* (September 8, 2011), at tinyurl.com/4ybawhx (both January 2012).

6. Big Blue

1. Jeff Kiger, "IBM, Roch. Rack Up Patent Records," *Rochester Post-Bulletin* (January 10, 2011), at tinyurl.com/5v24gwf and tinyurl.com/66hlfau (July 2011).

2. Not all observers agree with the suggestion that Rochester was a world unto itself. There are sharply polarized views, for instance, on the book by Frank Soltis, *Fortress Rochester*.

3. With the early development effort for the AS/400 system, eventually released in 1988, "the local [Rochester] development team of about 30 people reportedly operated without official funding and was free of interference from Big Blue's corporate office." Remembers one participant, "That was a fight all the way for Rochester. We spent more time fighting IBM than working with customers" (Jeff Kiger, "IBM of 1956 Is Much Different from IBM Today," *Rochester Post-Bulletin* [June 9, 2012], at tinyurl.com/c5mzpvw [July 2012]).

4. Timothy Prickett Morgan, "Rumored Layoffs at IBM Rochester Not True," *Four Hundred* newsletter 16:36 (September 17, 2007), at tinyurl.com/bp87njk (accessed July 2012).

5. The industry journal *Datamation* (July 1991) reported IBM's revenues from AS/400 at $14 billion, larger than DEC's $13.9 billion.

6. Jeff Kiger, "Roch. Native Named 'Person of Year,'" *Rochester Post-Bulletin* (June 3, 2011), at tinyurl.com/6bb5dj8 (July 2011).

7. Friedman, "John H. Patterson and the Sales Strategy of the National Cash Register Company, 1884 to 1922," 552–86.

8. Maney, *The Maverick and His Machine*, 155.

9. Howard Aiken is most likely the source; see Cohen, "Howard Aiken on the Number of Computers Needed for the Nation," 27–32.

10. See chapter 2 for ERA's contribution of magnetic-drum technology to the IBM 650.

11. See chapter 3 for a discussion of SAGE.

12. Minnesota Federal Writers' Project, *The WPA Guide to Minnesota*, 270 (pleasant little trade), 271 (cosmopolitan air), 370 (canning factory).

13. Bruce, *Eliot Noyes*, 153.

14. On Noyes, see ibid., 19, 39–185; Earls, *The Harvard Five in New Canaan*. John Harwood has recently published a major study of Noyes: *The Interface*.

15. Knowles and Leslie, "'Industrial Versailles,'" 1–33.

16. Saarinen in Martin, *The Organizational Complex*, 163. See also Pelkonen and Albrecht, eds., *Eero Saarinen*, 45–55.

17. "IBM Nickname 'Big Blue' Remains a Mystery," *Seattle Post-Intelligencer* (June 15, 2011), at tinyurl.com/6gl2g7d (July 2012); "No. 1's Awesome Strategy," *Business Week* (June 8, 1981), at tinyurl.com/6yga7y8 (July 2012).

18. Paul Rand's "eight-stripe" IBM logo of 1972 replaced an earlier thirteen-stripe logo used in the 1960s; see IBM Archives, at tinyurl.com/4u8j6 (August 2012). Gordon Bruce seems to be in error in attributing the eight-stripe logo to the year 1962 (*Eliot Noyes*, 152), though the fine print of a document indicates an official blue for IBM motor vehicles.

19. See IBM Archives, at tinyurl.com/5w4htxz (July 2012).
20. See the IBM public-relations FAQ, at tinyurl.com/5w9xkb4 (July 2012).
21. IBM Archives, "IBM 803 Proof Machine," at tinyurl.com/3uytu9b (July 2011).
22. Norberg and Yost, *IBM Rochester*, 17. As one programmer recalled, "you can work on a product, you can walk down a hall, and everybody that worked on it pretty much lives in that hallway, or in that building at least. And you knew everybody. And if you wanted to do something you just walk over and you talk. And it gave you a sense of being part of a team, contributing. I think management was very much pushing that type of philosophy too," stated Ben Persons and Herb Pelnar oral history, CBI OH 327, 32.
23. Timothy Prickett Morgan, "Rumored Layoffs at IBM Rochester Not True," *Four Hundred* newsletter, 16:36 (September 17, 2007), at tinyurl.com/bp87njk (accessed July 2012). Owing to exceptionally strong demand, prices on AS/400 models *increased* by an average of 26 percent from 1990 to 2000; see Jennifer Hagendorf, "AS/400 Price Hike To Spur Sales" (December 10, 1999), at tinyurl.com/5s6rz4h (July 2011).
24. See IBM Archives, "IBM System/3," at tinyurl.com/6gw7xzq (July 2011); Dick Hedger oral history, CBI OH 378, 7.
25. David L. Schleicher oral history, CBI OH 381, 9.
26. Bauer, Collar, and Tang, *The Silverlake Project*, 64.
27. System/38 had been "designed to come in and basically be the whole new basis for the Rochester product line," according to David L. Schleicher oral history, CBI OH 381, 7.
28. David L. Schleicher oral history, CBI OH 381, 11–14; Bauer, Collar, and Tang, *The Silverlake Project*, 25–57, 97.
29. David L. Schleicher oral history, CBI OH 381, 15–17; Bauer, Collar, and Tang, *The Silverlake Project*, 64–80.
30. AS/400 "cemented particularly the System/36 buyers confidence that Rochester can deliver. People were smart enough to realize that it was really System/36 running on a System/38 for that is what you've got going here, and new hardware tucked in the new system," noted David L. Schleicher oral history, CBI OH 381, 21, 34 (integrated database); Ben Persons and Herb Pelnar oral history, CBI OH 327, 14 (unique to Rochester).
31. David L. Schleicher oral history, CBI OH 381, 14, 34–37; Norberg and Yost, *IBM Rochester*, 34–38; Pine, "Design, Test, and Validation of the Application System/400 through Early User Involvement," 376 (tight integration).
32. Rosenblatt and Watson, "Concurrent Engineering," 22–37.
33. David L. Schleicher oral history, CBI OH 381, 17 (opening up the doors); Bauer, Collar, and Tang, *The Silverlake Project*, 133 (development team); Pine, "Design, Test, and Validation of the Application System/400 through Early User Involvement," 376–85.
34. David L. Schleicher oral history, CBI OH 381, 21 (within twelve hours); Bauer, Collar, and Tang, *The Silverlake Project*, 193 (ice cream). On IBM's Bruce Jawer, see tinyurl.com/6aypwp6 (July 2011). The Baldrige award was a personal goal of site general manager Larry Osterwise, according to Schleicher oral history, 63–65. The two-year effort required "salesmanship of the site that's trying to get it. I mean, first, you've got to sign up for it, get your people involved in it, as well as get Malcolm Baldrige people involved in it. And then, being able to present the information, know what to present, and to be able to get the ideas across," according to Ben Persons and Herb Pelnar oral history, CBI OH 327, 23.

35. Ben Persons and Herb Pelnar oral history, CBI OH 327, 34. IBM Rochester employed 4,400 in 2007, according to Timothy Prickett Morgan, "Rumored Layoffs at IBM Rochester Not True," *Four Hundred* newsletter 16:36 (September 17, 2007), available at tinyurl.com /bp87njk (accessed July 2012). "Big Blue no longer provides year-end reports of employee population at each campus. The last time it issued an official number, IBM reported that it had 4,200 employees in Rochester as of Dec. 31, 2008. In 2010, the company stopped releasing the number of workers it has in the United States," according to Jeff Kiger, "IBM Cuts Jobs This Week," *Rochester Post-Bulletin* (May 26, 2011), at tinyurl.com/3bnhmzo (July 2011). According to the annual *Minnesota Directory of Manufacturers,* IBM Rochester's employment was six thousand in 2001 and 5,100 in 2011.

36. Steve Lewis and Curt Mathiowetz oral history with author (June 5, 2008). "Since 1994 . . . Rochester's previous autonomy vanished. Rochester now is more of a job site for people outside of Rochester," suggested David L. Schleicher oral history, CBI OH 381, 70.

37. Laurie J. Flynn, "Technology Briefing: Intel to Ship Dual-Core Chips," *New York Times* (February 8, 2005), at tinyurl.com/6ducoh3 (July 2011). Intel abandoned the code-named Tejas, Nehalem, and Cedarmill chips, which had difficulties getting rid of up to 150 watts of heat, and instead developed Conroe as its first dual-core processor.

38. There were logical limits and financial constraints, however. "The implementation limit [for Blue Gene/L] is 128 racks. The largest actual implementation is 104. Blue Gene/P has an implementation limit of 256. The largest implementation is 40 today," stated Steve Lewis, oral history with author (June 5, 2008).

39. "Because they [LLNL] were going to buy so many racks, their money and payment for those racks, and the payment for research milestones up to the time of the racks, became a major piece of the business case. So consequently and as a result of their early commitment to the program with dollars, they got to influence the design," stated Lewis (ibid.).

40. See Almási et al., "An Overview of the Blue Gene/L System Software Organization," 543–55.

41. Steve Lewis and Curt Mathiowetz oral history with author (June 5, 2008).

42. Ibid.

43. Patrick Carey arranged for IBM Rochester to donate one of these Blue Gene development racks to the University of Minnesota, where in Keller Hall it is installed in a permanent exhibit. See Thomas J. Misa, "Director's Desk: IBM Blue Gene Supercomputer to CBI," *CBI Newsletter* 30:1 (spring 2008), and "Minnesota's Supercomputer," *CBI Newsletter* 32:2 (fall 2010), with a photo of the installation, available at www.cbi.umn.edu/about/newsletter.html. The Blue Gene exhibit panels are "Supercomputing Slows Down?" at tinyurl.com/bsu6xt9; "Minnesota Story," at tinyurl.com/bwgt6g3; and "Impact on Science and Technology," at tinyurl.com/cukvchn.

44. Curt Mathiowetz oral history with author (June 5, 2008).

45. On the Top500.org for November 2003, IBM's Watson Research Center is ranked seventy-third for a 1024-processor, or one-rack, Blue Gene/L DD1 Prototype (0.5GHz PowerPC 440). In June 2004, Rochester is ranked fourth for an 8192-processor, or eight-rack, Blue Gene/L DD1 Prototype. The November 2003 machine does not appear in IBM Rochester's historical listing at top500.org/site/history/2422 (July 2011).

46. LLNL's Blue Gene/L was ranked number one from June 2005 until June 2008, when it was ranked second, behind the IBM Roadrunner. In June 2011, LLNL's Blue Gene, now running 212,992 (dual-core) processors at 700 MHz, was ranked fourteenth.

47. Steve Lewis and Curt Mathiowetz oral history with author (June 5, 2008).

48. "The new #1 is DOE's IBM BlueGene/L beta system [with thirty-two racks] currently assembled and tested at the IBM Rochester site with a Linpack performance of 70.72 TFlop/s. This system, once completed, will be moved to the DOE's Lawrence Livermore National Laboratory" (top500.org/lists/2004/11). This thirty-two-rack beta system, while here credited to IBM Rochester, does not appear on the IBM Rochester site's listing, at top500.org/site/history/2422.

49. Steve Lewis oral history with author (June 5, 2008).

50. The loss of local culture and identity is strongly voiced in Kiger, "IBM of 1956 Is Much Different from IBM Today."

51. Dan Heilman, "IBM's Triple Play," *Twin Cities Business* (September 2008), at tinyurl.com/ctzt7qo; Jeff Kiger, "IBM's Smarter Game Show Goes Live Today," *Rochester Post-Bulletin* (February 14, 2011), at tinyurl.com/c2xf9hq; Jeff Kiger, "IBM's Watson and Jeopardy: The Rerun," *Rochester Post-Bulletin* (August 29, 2011), at tinyurl.com/3e86zb (all accessed July 2012).

52. See IBM Archives, "IBM Rochester: Harvest in the Heartland," at tinyurl.com/cdwybx and "Rochester Chronology," at tinyurl.com/86lksq (July 2012).

7. Industrial Dynamics

1. Dorfman, "Route 128," 303. See also Lewis, ed., *Manufacturing Suburbs*, 53–75; Lewis, *Chicago Made*.

2. For recent treatments, see Larson, *The White Pine Industry in Minnesota*; Register, *Packinghouse Daughter*; Ominsky, "Minnesota-Made Cars and Trucks," 93–112; Alanen, *Morgan Park*; "Historic St. Paul Ford Plant Closes," *Minneapolis Star Tribune* (December 17, 2011), at tinyurl.com/ccy9wjq (April 2013).

3. Elvery, "City Size and Skill Intensity," 367–79.

4. U.S. Bureau of Labor Statistics, *Occupational Wage Survey: Minneapolis–St. Paul, Minn.* (Washington, D.C.: Government Printing Office, 1954); BLS Bulletin No. 1172-5.

5. See Misa, ed., *Gender Codes*; and Ensmenger, *The Computer Boys Take Over*.

6. Minnesota Department of Employment Security, Research and Planning Section, *Occupational Outlook: The Minneapolis–St. Paul Metropolitan Area* (St. Paul: State of Minnesota, 1966), 16.

7. Ibid., 59.

8. Minnesota Department of Economic Development: Research Division, *The Business Machine Industry (SIC 357) in Minnesota and the U.S.: Market Performance, Trends and Projections* (St. Paul: Department of Economic Development, 1974).

9. Dorfman, "Route 128," 312. See also Lécuyer, *Making Silicon Valley*.

10. Cortada, *The Digital Hand*, 75; see also 38–41, 74–76, 107–10.

11. See Miller, ed., *The Rational Expectations Revolution*.

12. Gainor, "Into the 1990s"; Fettig, "Small Leak Teaches Big Lessons"; "Interview with Carl E. Powell [director of automation resources for the Federal Reserve System]," *Federal Reserve Bank of Minneapolis: The Region* (December 1991), available at tinyurl.com/evgxvlx (April 2013).

13. Josh Lowensohn, "Oregon Trail Facebook App to Be Replaced with Dating Service," CNET News (October 23, 2008), available at tinyurl.com/5ksjgl (accessed April 2013).

14. Dale LaFrenz oral history CBI OH 315, 5–6. LaFrenz notes his "U-Hi" math-teacher colleagues David C. Johnson, Pam Katzman, John Walther, Tom Kieren, and Larry Hatfield.

15. This is Robert Elijah Smith, author of *The Bases of FORTRAN*.

16. Univac 422 advertisement at tinyrul.com/bte66nm (April 2013). On Albrecht, see Levy, *Hackers*, 168–74.

17. John G. Kemeny, June 7, 1984, Princeton Mathematics Oral History Project, no. 22; John McCarthy, oral history interview, CBI OH 156. On Kemeny's Kids, see tinyurl.com/btq2odl (April 2013).

18. Dale LaFrenz oral history CBI OH 315, 8; Piele, "Computer Assisted Problem Solving in Mathematics."

19. Dale LaFrenz oral history CBI OH 315, 9.

20. Holznagel, Skow, and LaFrenz, "Historical Development of Minnesota's Instructional Computing Network," 79–80.

21. Dale LaFrenz oral history CBI OH 315, 15.

22. Sources include Haugo, "Minnesota Educational Computing Consortium"; Haugo, "MECC: A Management History," 152–58; Office of Technology Assessment, *Informational Technology and Its Impact on American Education* (Washington, D.C.: OTA, 1982), 214–21. There is a detailed timeline at tinyurl.com/cs4zua9 (April 2013). A full investigation of MECC, based on records held at the Minnesota Historical Society, is sorely needed.

23. Dale LaFrenz oral history CBI OH 315, 22 (giant computer); LaFrenz, "Planning for Instructional Time-Sharing Service," 214.

24. Brumbaugh, "Reflections on Educational Computing," 170; Minnesota Educational Computing Consortium, *1979–80 Microcomputing Report* (July 1979), appendix B, 36. Radio Shack and IBM were the other well-known bidders for the MECC contract. IBM's model 5110 at nearly twenty thousand dollars could not have bested the Apple offer, but Radio Shack's TRS-80, at $1,985, would have given Apple a run. Unaccountably, Radio Shack did not submit a timely bid.

25. Dale LaFrenz oral history CBI OH 315, 26.

26. "Atari and MECC Reach Agreement," *Infoworld* (November 1981): 2, 4.

27. For the origins of Oregon Trail, see Lussenhop, "Oregon Trail."

28. Minnesota Science and Technology Hall of Fame, at tinyurl.com/73rto9u; see also the approved biography at tinyurl.com/782jqfk (July 2012).

29. "Control Data Officer Starting Own Firm," *Los Angeles Times* (July 11, 1989), at tinyurl.com/bpq92qc (April 2013).

30. In the 1980s, France installed nine million Minitel terminals, which connected to the phone system and provided twenty-five million users a wealth of information; Minitel was shut down in July 2012. See Schofield, "Minitel."

31. Jean Polly acknowledged McCahill's priority: see tinyurl.com/cjosw40 (April 2013).

32. McCahill, "Partners in Gopherspace: The Next Frontier," March 1, 1991, in box 2, folder 9, McCahill papers; Frana, "Before the Web There Was Gopher." Alas, although in 2008 Firefox could read the Gopher-based history of Gopher at gopher://hal3000.cx, my current version returns "Firefox doesn't know how to open this address, because the protocol (gopher) isn't associated with any program." See the newsgroup archives at gopher.quux.org/1/Archives/ or install the Firefox Overbite FF extension to make the browser compatible with Gopher.

33. Lari, Buckeye, and Helbach, "Connecting Minnesota," 115; Reilly, "Trail of Woe Follows State Fiber Effort."

34. Mike Bollinger, Tech.MN news, "Midwest Internet Cooperative Exchange Forms to Speed Up Local Internet Traffic" (June 15, 2011), at tinyurl.com/brzdoyj (July 2012). MICE is at www.micemn.net.

8. High-Technology Innovation

1. Michael P. Moore, "The Genesis of Minnesota's Medical Alley," *University of Minnesota Medical Bulletin* (winter 1992), at tinyurl.com/99k9hd0; Minnesota Historical Society, "Oral History Collection: Pioneers of the Medical Device Industry in Minnesota, 1995–2001," at tinyurl.com/9y3m3fg. The Minneapolis garage where Medtronic was founded is profiled in Monica Smith, "Much More Than a Garage, a Place of Invention" (June 26, 2012), at tinyurl.com/8h4xsgs (all accessed September 2012).

2. Bloomberg News, "Medical Device Industry Helps Minnesota's Health," *Los Angeles Times* (January 28, 1997), at tinyurl.com/822jh3e (July 2012).

3. Minnesota Precision Manufacturing Association, "Manufacturing Fast Facts," at tinyurl.com/ckokwdm (April 2013).

4. See Isaacson and Young, *Under the Radar*.

5. On Benchmark, see *Minnesota Directory of Manufacturers* (MDM), 1980–2011 and company Web site at tinyurl.com/c8kks58.

6. Tata CEO Natarajan Chandrasekaran, quoted in "TCS Opens New Centre in Minneapolis," *Hindu Business Line* (September 18, 2012), at tinyurl.com/c5cpd3v; "TCS Inaugurates New Center in Minneapolis," *Consultant News* (September 20, 2012), at tinyurl.com/bl8uzfu (both September 2012).

7. Reuters profile at tinyurl.com/cspms48 (April 2013).

8. "NCR–Comten Adds Entry 5600 To Its 3745-Compatible Line," *Computer Business Review* (April 21, 1991), at tinyurl.com/6wg358r. See the "frozen" Comten Web site at ncrpr.ncr.com/support/comten/index.html (April 2013).

9. Bermo: *Minnesota Directory of Manufacturers*, 1980–2011 and company Web site at www.bermo.com.

10. Metalcraft: *Minnesota Directory of Manufacturers*, 1990–2011 and company Web site at www.metal-craft.com.

11. Hutchinson: *Minnesota Directory of Manufacturers*, 1980–2011 and company Web site at www.htch.com.

12. "New CEO's Experience Spans More Than 20 Years in Industry," *Intercom: A Newsmagazine for Memorex People* 17:1 (February 1980): 3, at mrxhist.org/docs/Bede_5301.pdf.

13. Graeme Thickins, "Bloomington Firm Expects to Add 300 High-Tech Jobs," Minnov8.com (November 11, 2010), available at tinyurl.com/89rxu2l (July 2012).

14. See ww3.startribune.com/projects/st100/employeeView.php (April 2013).

15. Mark Reilly, "Tata's Minnesota Growth a Sign of U.S. Hiring by India Firms," *Minneapolis/St. Paul Business Journal* (August 28, 2012), at tinyurl.com/9jnhwkd.

16. Peter C. Patton oral history, CBI OH 325, 61.

17. Dee DePass, "St. Paul-Based Lawson Software Hands Out Pink Slips," *Star-Tribune* (July 13, 2011), at tinyurl.com/8pa2hy6; "Lawson Software Agrees to $2 Billion Buyout," *New York Times* (April 26, 2011), at tinyurl.com/9zqthzk (both July 2012).

18. Dayton and Tata CEO Natarajan Chandrasekaran, quoted in "TCS Opens New Centre in Minneapolis," *Hindu Business Line* (September 18, 2012), at tinyurl.com/c5cpd3v; "TCS Inaugurates New Center in Minneapolis," *Consultant News* (September 20, 2012), at tinyurl.com/bl8uzfu (both September 2012).

19. Brandon Shackelford, "Businesses Concentrate Their R&D in a Small Number of Geographic Areas in the United States," National Center for Science and Engineering Statistics InfoBrief in NSF 12-326 (September 2012); Minnesota R&D 2012, at tinyurl.com/9fn6bod; State Government Research and Development: Fiscal Year 2009: Detailed Statistical Tables in NSF 12-331 (September 2012).

20. Los Alamos definition of nanotechnology at tinyurl.com/cjsfgz3 (April 2013).

21. "The 2007 Nobel Prize in Physics: Press Release," at tinyurl.com/d52kxcr (April 2013).

22. See the Physics and Nanotechnology Building at physicsnano.umn.edu (April 2013).

23. One apposite instance of the connections between microelectronics and nanotechnology can be seen in the textbooks authored by University of Minnesota professor Stephen A. Campbell, director of the Nanofabrication Center. His books' 1996 and 2001 editions from Oxford University Press are titled *The Science and Engineering of Microelectronic Fabrication*, while the title becomes *Fabrication Engineering at the Micro and Nanoscale* for the third (2008) and fourth (2012) editions.

Appendix

1. Chamber of Commerce, at tinyurl.com/bqs77wz (July 2012).

2. 2009 figure MDM.

3. Ball Electronic [1980] and Bermo [1990–] have same address.

4. 2005 figure from MDM (nearest available).

5. CDC acquired Autocon Industries; joint venture Magnetic Data Inc. (neither separately listed).

6. Frank Dawe memo, May 6, 1990, in CBI (80), series 9, box 30, folder 18: WCN day files. 1989 figure for 1990.

7. Control Data Systems, Inc. (Arden Hills) became Syntegra; Peripheral products bought by Seagate.

8. 1999 figure MDM: Cray Research listed as Silicon Graphics (Eagan).

9. St. Paul move 2009, www.startribune.com/local/east/51342087.html.

10. Data from government-contractor.bizdirlib.com/ceo/DataSource_Pomeroy.

11. 1999 figure MDM.

12. 2 plants 2010 MDM figures.

13. 1999 figure MDM.

14. 1980 figure from IBM Archives, at tinyurl.com/7kr7hvk.

15. 1991 figure from IBM Archives, at tinyurl.com/86lksqk.

16. 2007 figure from MDM.

17. 1983 figure from www.inc.com/inc5000/profile/lawson-associates.

18. From www.lawson.com/About-Lawson/Company-Overview/History/.

19. National Electronics [1990] and Micro Dynamics [2001–] shared town/zip/phone numbers.

20. 2008 figure MDM.

21. DataMyte business of Rockwell Automation was originally incorporated in 1965 as Electro General Corp., www.cbi.umn.edu/resources/mncomphist-d.html.

22. 1998 figure of Seagate's 4,500 Minnesota jobs from Dirk Deyoung, "Shakopee Lands Seagate Division: 1,200 Employees to Relocate from Bloomington," *Minneapolis/St. Paul Business Journal* (January 4, 1998), available at tinyurl.com/7enmcyz.

23. Seagate 2001: Shakopee 1,000, Bloomington 3,000.

24. 2007 MDM lists *only* 7801 Computer Ave. at 1,500 jobs. Seagate not listed in recent MDM.

25. 2009 figure MDM.

26. 1999 figure MDM.

27. Successor to Control Data Systems Inc. (Arden Hills).

28. 2002 listing in MDM appears to be the last.

29. Telex acquired by (German) Bosch Group in 2006.

30. 2007 figure MDM Sagebrush (Minneapolis).

Bibliography

Archives and Manuscripts

Charles W. Bradley Collection on the ENIAC Trial (CBI 140), Charles Babbage Institute, University of Minnesota, Minneapolis.

I. Bernard Cohen Papers (CBI 182), Charles Babbage Institute, University of Minnesota, Minneapolis.

Control Data Corporation Collection, (CBI 80), Charles Babbage Institute, University of Minnesota, Minneapolis.
- Acquisitions, Subsidiaries, and Joint Ventures
- Executive Papers, 1956–91
- Executive History Project Records (1980), 1957–81
- Executive Papers —William C. Norris, 1946–91
- Facilities, 1959–86
- News Releases
- Research and Development Project Reports

Engineering Research Associates (ERA)—Remington Rand—Sperry Rand Records (CBI 176), Charles Babbage Institute, University of Minnesota, Minneapolis.

ENIAC Patent Trial Collection, 1864–1973, University Archives and Records Center of the University of Pennsylvania; at tinyurl.com/2e86l.

ENIAC Trial Exhibits Master Collection (CBI 145), Charles Babbage Institute, University of Minnesota, Minneapolis.

Honeywell, Inc., Honeywell vs. Sperry Rand Records (CBI 1), Charles Babbage Institute, University of Minnesota, Minneapolis.

Earl R. Larson Papers (CBI 31), Charles Babbage Institute, University of Minnesota, Minneapolis.

Robert V. Leacock Collection of ERA and Univac scrapbooks (CBI 57), Charles Babbage Institute, University of Minnesota, Minneapolis.

Mark P. McCahill Papers (CBI 195), Charles Babbage Institute, University of Minnesota, Minneapolis.

John E. Parker Papers (CBI 100), Charles Babbage Institute, University of Minnesota, Minneapolis.

Andrew Russell (Drew) Pearson collection, American University Archives and Special Collections, box 9, folder 20, available at tinyurl.com/3l849e3 (March 2011).

Sperry Rand Corporation, Univac Division. Honeywell vs. Sperry Litigation Records (CBI 72), Charles Babbage Institute, University of Minnesota, Minneapolis.

Sperry–UNIVAC Company Records, Hagley Museum and Library; at tinyurl.com/7yj6kgj.

Marvin L. Stein Papers (UARC 1140). University Archives, University of Minnesota, Minneapolis.

Frank M. Verzuh, Moore School of Electrical Engineering lecture notes (CBI 51), Charles Babbage Institute, University of Minnesota, Minneapolis.

Oral Histories

All oral histories, unless otherwise noted, are available online at www.cbi.umn.edu/oh.

Allard, Raymond W. Oral history interview by Griff Kennedy (June 21, 1982), Charles Babbage Institute OH 288, [n.p.].

Anderson, Walter L. Oral history interview by Arthur L. Norberg (September 11, 1986). Charles Babbage Institute OH 119.

Auerbach, Isaac L. Oral history interview by Nancy B. Stern (April 10, 1978), Charles Babbage Institute OH 2.

Baker, Eugene L. Oral history interview by William H. Fuhr (February 14, 1980), Charles Babbage Institute OH 293 [n.p.] (not online).

Brown, Gordon R. Oral history interview by William H. Fuhr (February 14, 1980), Charles Babbage Institute OH 294 [n.p.] (not online).

Bulver, Paul J. Oral history interview by Griff Kennedy (June 2, 1982), Charles Babbage Institute OH 279 [n.p.].

Bulver, Paul J. Oral history interview by William H. Fuhr (February 12, 1980), Charles Babbage Institute OH 300 [n.p.].

Campaigne, Howard. Oral history interview by Robert D. Farley (June 28, 1983), National Security Agency NSA-OH-14-83. Available at tinyurl.com/6d7dqgg (June 2011).

Chambers, Carl. Oral history interview by Nancy B. Stern (November 30, 1977), Charles Babbage Institute OH 7.

Clover, H. Dick. Oral history interview by Arthur L. Norberg (June 5, 1986), Charles Babbage Institute OH 113.

Cohen, Arnold A. Oral history interview by Arthur L. Norberg (July 2, 1987), Charles Babbage Institute OH 138.

Cohen, Arnold A. Oral history interview by James Baker Ross (January 20 and 28, February 9 and 16, and March 2 and 9, 1983), Charles Babbage Institute OH 58.

Control Data Corporation. Oral history interview moderated by Neil R. Lincoln (May and September 1975), Charles Babbage Institute OH 321.

Cray, Seymour. Oral history interview by David Allison, Smithsonian Institution (May 9, 1995); at tinyurl.com/bpbowzm (April 2013).

Crichton, Charles F. Oral history interview by Mollie Price (October 7, 1981), Charles Babbage Institute OH 264, [n.p.] (not online).

Davidov, Marv. Oral history by Clarke A. Chambers (September 27, 1995), University of Minnesota Archives, purl.umn.edu/49116 (August 2011).

Drake, Willis K. Oral history interview by James Baker Ross (February 3, 1983), Charles Babbage Institute OH 46.

Eckert, J. Presper. Oral history interview by Nancy B. Stern (October 28, 1977), Charles Babbage Institute OH 13.

Fein, Louis. Oral history interview by Pamela McCorduck (May 9, 1984), Charles Babbage Institute OH 15.

Forrest, Henry S. Oral history interview by Mollie Price (December 6, 1982), Charles Babbage Institute OH 289.

Grabbe, Dimitry. Oral history interview by Robert Colburn (December 18, 2007), IEEE History Center #476, tinyurl.com/3c3nkjy (July 2011).

Gunderson, Richard C. Oral history interview by Mollie Price (October 23, 1981), Charles Babbage Institute OH 266.

Harris, James D. Oral history interview by Mollie Price (September 1981), Charles Babbage Institute OH 295.

Hedger, Dick. Oral history interview by Philip Frana (May 17, 2001), Charles Babbage Institute OH 378.

Hill, John L. Oral history interview by Arthur L. Norberg (January 15 and 22, 1986), Charles Babbage Institute OH 101.

Holberton, Frances E. Oral history interview by James Baker Ross (April 14, 1983), Charles Babbage Institute OH 50.

Kamp, Thomas G. Oral history interview by William H. Fuhr (February 25, 1980), Charles Babbage Institute OH 297 [n.p.] (not online).

Kemeny, John G. Oral history interview by Albert Tucker (June 7, 1984), Princeton Mathematics Oral History Project, no. 22.

LaFrenz, Dale Eugene. Oral history interview by Judy E. O'Neill (April 13, 1995), Charles Babbage Institute OH 315.

Lareau, Richard G. Oral history interview by Mollie Price (April 16, 1982), Charles Babbage Institute OH 273.

Lewis, Steve, and Curt Mathiowetz. Oral history interview with author (June 5, 2008).

McCarthy, John. Oral history interview by William Aspray (March 2, 1989), Charles Babbage Institute OH 156.

McDonald, Robert Emmett. Oral history interview by James Baker Ross (December 16, 1982), Charles Babbage Institute OH 45.

Mullaney, Frank C. Oral history interview by Arthur L. Norberg (June 2 and 11, 1986), Charles Babbage Institute OH 110.

Mumma, Robert E. Oral history interview by William Aspray (April 19, 1984), Charles Babbage Institute OH 73.

Norris, William C. Oral history interview by Arthur L. Norberg (July 28 and October 1, 1986), Charles Babbage Institute OH 116.

Parker, John E. Oral history interview by Arthur L. Norberg (December 13, 1985, and May 6, 1986), Charles Babbage Institute OH 99.

Patton, Peter C. Oral history interview by Philip L. Frana (August 30, 2000), Charles Babbage Institute OH 325.

Pendergrass, James T. Oral history interview by William Aspray (March 28, 1985), Charles Babbage Institute OH 93.

Persons, Ben, and Herb Pelnar. Oral history interview by Philip L. Frana (July 17, 2001), Charles Babbage Institute OH 327.

Price, Robert M. Oral history interview with author (March–July 2009). Published as Misa, ed., *Building the Control Data Legacy.*

Rench, Carl F. Oral history interview by William Aspray (April 18, 1984), Charles Babbage Institute OH 72.

Robinson, Herbert W. Oral history interview by Bruce Bruemmer (July 13, 1988), Charles Babbage Institute OH 147.

Rubens, Sidney Michel. Oral history interview by Arthur L. Norberg (January 6 and 15, 1986), Charles Babbage Institute OH 100.

Ryden, Arnold J. Oral history interview by Judy E. O'Neill (April 5, 1995), Charles Babbage Institute OH 314.

Schleicher, David L. Oral history interview by Arthur L. Norberg (January 24, 2006), Charles Babbage Institute OH 381.

Schumacher, Mike. Oral history interview by Griff Kennedy (June 23, 1982), Charles Babbage Institute OH 287 (not online).

Slattery, Leo F. Oral history interview by Griff Kennedy (June 4, 1982), Charles Babbage Institute OH 284 [n.p.].

Stein, Marvin L. Oral history interview by William Aspray (October 29 and November 7, 1984), Charles Babbage Institute OH 90.

Thorndyke, Lloyd M. Oral history interview by Griff Kennedy (June 4, 1982), Charles Babbage Institute OH 280 (not online).

Thorndyke, Lloyd M. Oral history interview by William H. Fuhr (February 18, 1980), Charles Babbage Institute OH 303 [n.p.].

Trousdale, Elmer B. Oral history interview by Dan Norris and Debra Fischer (circa 1977), Charles Babbage Institute OH 271 [n.p.] (not online).

UNIVAC Conference. Oral history on May 17–18, 1990. Charles Babbage Institute OH 200.

Zemlin, Richard A. Oral history interview by Bruce Bruemmer (May 16, 1988), Charles Babbage Institute OH 152.

Zemlin, Richard A. Oral history interview by Mollie Price (June 21, 1983), Charles Babbage Institute OH 307 (not online).

Published Works

Agar, Jon. *The Government Machine: A Revolutionary History of the Computer.* Cambridge, Mass.: MIT Press, 2003.

Alanen, Arnold R. *Morgan Park: Duluth, U.S. Steel, and the Forging of a Company Town.* Minneapolis: University of Minnesota Press, 2007.

Almási, George, et al. "An Overview of the Blue Gene/L System Software Organization." *Lecture Notes in Computer Science* 2790 (2003): 543–55. At tinyurl.com/3v98oqk (July 2011).

Anderson, George P. "Failure Modes in High Reliability Components." *Transactions of the American Society for Quality Control 1962* (Cincinnati, May 1962): 353–62. Available at asq.org/qic/display-item/?item=689 (September 2012).

Anderson, George P., and L. E. Peterson (Remington Rand Univac). "Analysis of Random Failures." *Planetary and Space Science* 7 (July 1961): 242–53. At tinyurl.com/3dp35sh.

Anderson, Jack. *Confessions of a Muckraker: The Inside Story of Life in Washington during the Truman, Eisenhower, Kennedy and Johnson Years.* New York: Random House, 1979.

Aris, Rutherford. *The Optimal Design of Chemical Reactors: A Study in Dynamic Programming.* New York and London: Academic Press, 1961.

Aspray, William. *John von Neumann and the Origins of Modern Computing.* Cambridge, Mass.: MIT Press, 1990.

Atkinson, Paul. "The Curious Case of the Kitchen Computer: Products and Non-Products in Design History." *Journal of Design History* 23:2 (2010): 163–79.

Bauer, Roy A., Emilio Collar, and Victor Tang. *The Silverlake Project: Transformation at IBM.* New York: Oxford University Press, 1992.

Baum, Claude. *The System Builders: The Story of SDC.* Santa Monica, Calif.: System Development Corporation, 1981.

Beardsell, Mark, and Vernon Henderson. "Spatial Evolution of the Computer Industry in the USA." *European Economic Review* 43 (1999): 431–56.

Belussi, Fiorenza, and Katia Caldari. "At the Origin of the Industrial District: Alfred Marshall and the Cambridge School." *Cambridge Journal of Economics* 33:2 (2009): 335–55.

Berlin, Leslie. *The Man behind the Microchip: Robert Noyce and the Invention of Silicon Valley.* New York: Oxford University Press, 2005.

Bernstein, Barton J. "Reconsidering the 'Atomic General': Leslie R. Groves." *Journal of Military History* 67:3 (2003): 883–920.

Blackman, Nelson M. *A Survey of Automatic Digital Computers.* Washington, D.C.: Office of Naval Research, 1953.

Boslaugh, David L. *When Computers Went to Sea: The Digitization of the United States Navy.* Los Alamitos, Calif.: IEEE Computer Society Press, 1999.

Britcher, Robert N. *The Limits of Software.* Reading, Mass.: Addison-Wesley, 1999.

Brockman, Vicky Lynn. "Social Movement Repertoires and Dynamics: A Study of the Honeywell Project and WAMM." University of Minnesota, Ph.D. thesis, 1998.

Bruce, Gordon. *Eliot Noyes.* London/New York: Phaidon, 2006.

Brumbaugh, Kenneth. "Reflections on Educational Computing." *Creative Computing* 10:11 (November 1984): 170.

Brusehaver, Tom. "Linux in Air Traffic Control." *Linux Journal* (January 1, 2004); at tinyurl.com/3jltchp (May 2011).

Brusentsov, Nikolay Petrovich, and José Ramil Alvarez. "Ternary Computers: The Setun and the Setun 70." In J. Impagliazzo and E. Proydakov, eds., SoRuCom 2006, *IFIP Advances in Information and Communication Technology,* vol. 357 (2011), 74–80.

Buchholz, Werner, ed. *Planning a Computer System: Project Stretch.* New York: McGraw-Hill, 1962.

Burke, Colin. "An Introduction to a Historic Computer Document: Betting on the Future—The 1946 Pendergrass Report Cryptanalysis and the Digital Computer." *Cryptologic Quarterly* 13:4 (winter 1994): 65–75. At tinyurl.com/3vt659b (May 2011).

———. *Information and Secrecy: Vannevar Bush, Ultra, and the Other Memex.* Metuchen, N.J.: Scarecrow Press, 1994.

Burks, Alice Rowe. *Who Invented the Computer?: The Legal Battle That Changed Computing History.* New York: Prometheus Books, 2003.

Burks, Alice R., and Arthur W. Burks. *The First Electronic Computer: The Atanasoff Story.* Ann Arbor: University of Michigan Press, 1988.

Burns, Thomas L. "The Origins of the National Security Agency 1940–1952." Fort Meade, Md.: National Security Agency Center for Cryptologic History, 1990.

Campbell, S. G., P. S. Herwitz, and J. H. Pomerene. "A Nonarithmetical System Extension." In Werner Buccholz, ed., *Planning a Computer System: Project Stretch.* New York: McGraw-Hill, 1962, 254–72.

Campbell, Stephen A. *Fabrication Engineering at the Micro and Nanoscale.* New York: Oxford University Press, 2008.

———. *The Science and Engineering of Microelectronic Fabrication.* New York: Oxford University Press, 1996.

Campbell-Kelly, Martin, and Michael Williams, eds. *The Moore School Lectures.* Cambridge, Mass.: MIT Press, 1985; CBI–Tomash reprint series volume 9.

Ceruzzi, Paul. *A History of Modern Computing.* Cambridge, Mass.: MIT Press, 1998.

———. *Internet Alley: High Technology in Tysons Corner, 1945–2005.* Cambridge, Mass.: MIT Press, 2008.

———. "'The Mind's Eye' and the Computers of Seymour Cray." In Bernard Finn, ed., *Exposing Electronics.* Amsterdam: Harwood Academic Publishers, 2000, 151–60.

———. *Reckoners: The Prehistory of the Digital Computer, from Relays to the Stored Program Concept, 1935–1945.* Westport, Conn.: Greenwood Press, 1983.

Chapin, George G. "Organizing and Programming a Shipboard Real-Time Computer System." *Fall Joint Computer Conference Proceedings* (1963): 127–37.

Cohen, I. Bernard. "Howard Aiken on the Number of Computers Needed for the Nation." *IEEE Annals of the History of Computing* 20:3 (1998): 27–32.

Copeland, B. Jack, et al. *Colossus: The Secrets of Bletchley Park's Codebreaking Computers.* Oxford: Oxford University Press, 2006.

Cortada, James W. *Before the Computer: IBM, NCR, Burroughs, and Remington Rand and the Industry They Created, 1865–1956.* Princeton, N.J.: Princeton University Press, 1993.

———. *The Digital Hand: Volume 2: How Computers Changed the Work of American Financial, Telecommunications, Media, and Entertainment Industries.* New York: Oxford University Press, 2006.

DeBrosse, Jim, and Colin Burke. *The Secret in Building 26: The Untold Story of America's Ultra War against the U-boat Enigma Codes.* New York: Random House, 2004.

Diers, John W., and Aaron Isaacs. *Twin Cities by Trolley: The Streetcar Era in Minneapolis and St. Paul.* Minneapolis: University of Minnesota Press, 2007.

Dorfman, Nancy S. "Route 128: The Development of a Regional High Technology Economy." *Research Policy* 12 (1983): 299–316.

Drake, Willis K. "A Look at the Local Scene." *Upper Midwest Investor* 1:1 (April 1961): 6–7.

Dyson, George. *Turing's Cathedral: The Origins of the Digital Universe.* New York: Pantheon, 2012.

Earls, William D. *The Harvard Five in New Canaan: Midcentury Modern Houses by Marcel Breuer, Landis Gores, John Johansen, Philip Johnson, Eliot Noyes.* New York: Norton, 2006.

Elvery, Joel A. "City Size and Skill Intensity." *Regional Science and Urban Economics* 40:6 (November 2010): 367–79.

Engineering Research Associates. *High-Speed Computing Devices.* New York: McGraw-Hill, 1950; CBI–Tomash reprint series volume 4.

Ensmenger, Nathan. *The Computer Boys Take Over: Computers, Programmers, and the Politics of Technical Expertise.* Cambridge, Mass.: MIT Press, 2010.

Fettig, David. "Small Leak Teaches Big Lessons." *Federal Reserve Bank of Minneapolis: The Region* (June 1991). At tinyurl.com/c27brz2 (April 2013).

Frana, Philip. "Before the Web There Was Gopher." *IEEE Annals of the History of Computing* 26:1 (2004): 20–41.

Friedman, Walter A. "John H. Patterson and the Sales Strategy of the National Cash Register Company, 1884 to 1922." *Business History Review* (winter 1999): 552–86.

Fritz, W. Barkley. "The Women of ENIAC." *IEEE Annals of the History of Computing* 18:3 (1996): 13–28.

Gainor, Thomas E. "Into the 1990s: Continuing Challenges for Payments Systems." *Federal Reserve Bank of Minneapolis: The Region* (August 1989). At tinyurl.com/cbwud3a (April 2013).

General Accounting Office. *Air Traffic Control: Software Problems at Control Centers Need Immediate Attention.* Washington, D.C.: U.S. GAO, 1991; IMTEC-92-1.

Gibson, David V., and Everett M. Rogers. *R&D Collaboration on Trial: The Microelectronics and Computer Technology Corporation.* Boston: Harvard Business School Press, 1994.

Goldstine, Herman H. *The Computer from Pascal to von Neumann.* Princeton, N.J.: Princeton University Press, 1972.

Government Accountability Office. *National Airspace System: FAA Has Made Progress but Continues to Face Challenges in Acquiring Major Air Traffic Control Systems.* Washington, D.C.: U.S. GAO, 2005.

Graf, R. W. *Case Study of the Development of the Naval Tactical Data System.* National Academy of Sciences, Committee on the Utilization of Scientific and Engineering Manpower, 1964.

Gray, George. "Engineering Research Associates and the Atlas Computer." *Unisys History Newsletter* (June 1999). At tinyurl.com/4xuuuvy.

Groves, Leslie R. *Now It Can Be Told: The Story of the Manhattan Project.* New York: Harper, 1962.

Green, Thomas J. *Bright Boys.* Natick, Mass.: A. K. Peters, 2010.

Halmos, P. R. "The Legend of John von Neumann." *American Mathematical Monthly* 80:4 (April 1973): 382–94. At www.jstor.org/stable/2319080.

Hamming, Richard W. *Coding and Information Theory.* 2d ed. Englewood Cliffs, N.J.: Prentice-Hall, 1986.

———. "Error Detecting and Error Correcting Codes." *Bell System Technical Journal* 29:2 (1950): 147–60.

Hart, David M. "From 'Ward of the State' to 'Revolutionary without a Movement': The Political Development of William C. Norris and Control Data Corporation, 1957–1986." *Enterprise and Society* 6:2 (June 2005): 197–223.

Harwood, John. *The Interface: IBM and the Transformation of Corporate Design, 1945–1976.* Minneapolis: University of Minnesota Press, 2011.

Haugo, John E. "MECC: A Management History." In *Technology and Education: Policy, Implementation, Evaluation: Proceedings of the National Conference on Technology and Education,* January 26–28, 1981. Washington, D.C.: Institute for Educational Leadership, 1981, 152–58.

———. "Minnesota Educational Computing Consortium." Paper presented at the Association for Educational Data Systems Annual Convention, New Orleans, April 16–19, 1973 (ERIC ED 087 434).

Hayes, Brian. "Third Base." *American Scientist* 89:6 (November–December 2001): 490.

Heims, Steve J. *John von Neumann and Norbert Wiener: From Mathematics to the Technologies of Life and Death.* Cambridge, Mass.: MIT Press, 1980.

Hodges, Andrew. *Alan Turing: The Enigma.* New York: Simon and Schuster, 1983.

Holbrook, Daniel. "Controlling Contamination: The Origins of Clean Room Technology." *History and Technology* 25:3 (2009): 173–91.

Holznagel, Donald, Mike Skow, and Dale LaFrenz. "Historical Development of Minnesota's Instructional Computing Network." In *Proceedings of the ACM 1975 Annual Conference.* New York: ACM, 1975, 79–80.

Hooks, Gregory. *Forging the Military–Industrial Complex: World War II's Battle of the Potomac.* Urbana: University of Illinois Press, 1991.

House, Charles H., and Raymond L. Price. *The HP Phenomenon: Innovation and Business Transformation.* Stanford, Calif.: Stanford Business Books, 2009.

Hsu, David H., and Martin Kenney. "Organizing Venture Capital: The Rise and Demise of American Research & Development Corporation, 1946–1973." *Industrial and Corporate Change* 14:4 (August 2005): 579–616.

Hughes, Thomas P. *Elmer Sperry: Inventor and Engineer.* Baltimore: Johns Hopkins University Press, 1971.

Irvine, M. M. "Early Digital Computers at Bell Telephone Laboratories." *IEEE Annals of the History of Computing* 23:3 (2001): 22–42.

Isaacson, Bob, and Neal Young. *Under the Radar: Minnesota's Defense Industry.* St. Paul: Department of Employment and Economic Development, 2006.

Israel, David R. "The Application of a High-Speed Digital Computer to the Present-Day Air Traffic Control System." Cambridge, Mass.: MIT Digital Computer Laboratory Report R-203, January 15, 1952.

Jacobs, John F. *The SAGE Air Defense Systems: A Personal History.* Bedford, Mass.: MITRE Corporation, 1986.

Kahn, David. *Codebreakers.* New York: Macmillan, 1967.

Knowles, Scott G., and Stuart W. Leslie. "'Industrial Versailles': Eero Saarinen's Corporate Campuses for GM, IBM, and AT&T." *Isis* 92:1 (2001): 1–33.

LaFrenz, Dale. "Planning for Instructional Time-Sharing Service." SIGUCCS '77 Proceedings of the Fifth Annual ACM SIGUCCS Conference on User Services. New York: ACM, 1977, 126.

Lampe, David. *The Massachusetts Miracle: High Technology and Economic Revitalization.* Cambridge, Mass.: MIT Press, 1988.

Lari, Adeel Z., Kenneth R. Buckeye, and Mary L. Helbach. "Connecting Minnesota: Superhighways to Information Superhighways." *Transportation Research Record: Journal of the Transportation Research Board* 1690 (1999): 114–20.

Larson, Agnes M. *The White Pine Industry in Minnesota: A History.* Minneapolis: University of Minnesota Press, 2007.

Layton, Edwin T. *And I Was There: Pearl Harbor and Midway—Breaking the Secrets.* New York: William Morrow, 1985.

Lécuyer, Christophe. *Making Silicon Valley: Innovation and the Growth of High Tech, 1930–1970.* Cambridge, Mass.: MIT Press, 2006.

Lee, John A. N., Colin Burke, and Deborah Anderson. "The US Bombes, NCR, Joseph Desch, and 600 WAVES: The First Reunion of the US Naval Computing Machine Laboratory." *IEEE Annals of the History of Computing* 22:3 (2000): 27–41.

Lemaitre, G., and M. S. Vallarta. "On the Geomagnetic Analysis of Cosmic Radiation." *Physical Review* 49 (1936): 719–26.

Levy, Steven. *Hackers.* Garden City, N.Y.: Doubleday, 1984.

Lewis, Robert. *Chicago Made: Factory Networks in the Industrial Metropolis.* Chicago: University of Chicago Press, 2008.

———, ed. *Manufacturing Suburbs: Building Work and Home on the Metropolitan Fringe.* Philadelphia: Temple University Press, 2004.

Lewis, Sinclair. *Babbitt.* New York: Harcourt, Brace and Company, 1922.

Light, Jennifer S. "When Computers Were Women." *Technology and Culture* 40:3 (1999): 455–83.

Lojek, Bo. *History of Semiconductor Engineering.* Berlin: Springer Verlag, 2007.

Lundstrom, David E. *A Few Good Men from Univac.* Cambridge, Mass.: MIT Press, 1987.

Lussenhop, Jessica. "Oregon Trail: How Three Minnesotans Forged Its Path." *City Pages* (January 19, 2011). At tinyurl.com/cvlk4qt (April 2013).

MacKenzie, Donald. *Knowing Machines.* Cambridge, Mass.: MIT Press, 1996.

Maney, Kevin. *The Maverick and His Machine: Thomas Watson, Sr., and the Making of IBM.* Hoboken, N.J.: John Wiley, 2003.

Markusen, Ann. "Sticky Places in Slippery Space: A Typology of Industrial Districts." *Economic Geography* 72:3 (1996): 293–313.

Marshall, Alfred. *Principles of Economics* (1890). At www.econlib.org/library/Marshall/marP.html.

Martin, Reinhold. *The Organizational Complex: Architecture, Media, and Corporate Space.* Cambridge, Mass.: MIT Press, 2003.

Masters, Carol, and Marv Davidov. *You Can't Do That!: Marv Davidov, Nonviolent Revolutionary.* Minneapolis: Nodin Press, 2009.

McCartney, Scott. *ENIAC: the Triumphs and Tragedies of the World's First Computer.* New York: Walker, 1999.

McTiernan, Charles E. "The ENIAC Patent." *IEEE Annals of the History of Computing* 20:2 (1998): 54–58, 80.

Miller, Preston J., ed. *The Rational Expectations Revolution: Readings from the Front Line.* Cambridge, Mass.: MIT Press, 1994.

Millett, Larry. *AIA Guide to the Twin Cities.* St. Paul: Minnesota Historical Society Press, 2007.

Milward, Alan. *War, Economy, and Society, 1939–1945.* Berkeley: University of California Press, 1977.

Minnesota Federal Writers' Project. *The WPA Guide to Minnesota.* St. Paul: Minnesota Historical Society Press, 1985; reprinted from original 1938.

Misa, Thomas J., ed. *Building the Control Data Legacy: The Career of Robert M. Price.* Minneapolis: Charles Babbage Institute, 2012.

———. *Gender Codes: Why Women Are Leaving Computing.* Hoboken, N.J.: Wiley/IEEE Computer Society Press, 2010.

Misa, Thomas J., and Robert W. Seidel, eds. *College of Science and Engineering: The Institute of Technology Years (1935–2010)*. Minneapolis: Charles Babbage Institute, 2010.

Mozingo, Louise A. *Pastoral Capitalism: A History of Suburban Corporate Landscapes.* Cambridge, Mass.: MIT Press, 2011.

Murray, Charles J. *The Supermen: The Story of Seymour Cray and the Technical Wizards behind the Supercomputer.* New York: John Wiley, 1997.

Nessell, C. W. *Honeywell: The Early Years.* Minneapolis: Minneapolis-Honeywell Regulator Company, 1960.

———. *The Restless Spirit.* Minneapolis: Minneapolis-Honeywell Regulator Company, 1963.

Norberg, Arthur L. *Computers and Commerce: A Study of Technology and Management at Eckert-Mauchly Computer Company, Engineering Research Associates, and Remington Rand, 1946–1957.* Cambridge, Mass.: MIT Press, 2005.

Norberg, Arthur L., and Jeffrey R. Yost. *IBM Rochester: A Half Century of Innovation.* Rochester, Minn.: IBM, 2006.

Norberg, Arthur L., and Judy E. O'Neill. *Transforming Computer Technology: Information Processing for the Pentagon, 1962–1986.* Baltimore: Johns Hopkins University Press, 1996.

Norris, Robert S. *Racing for the Bomb: General Leslie R. Groves, the Manhattan Project's Indispensable Man.* South Royalton, Vt.: Steerforth Press, 2002.

Ominsky, Alan. "Minnesota-Made Cars and Trucks." *Minnesota History* (fall 1972): 93–112.

Owens, Larry. "Vannevar Bush and the Differential Analyzer: The Text and Context of an Early Computer." *Technology and Culture* 27 (1986): 63–95.

Paolucci, Dominic Anthony, Norman C. Polmar, and John Patrick. *A Guide to U.S. Navy Command, Control, and Communications.* San Diego, Calif.: Naval Ocean Systems Center, 1979; Naval Ocean Systems Center Technical Document 247.

Parker, Frederick D. *A Priceless Advantage: U.S. Navy Communications Intelligence and the Battles of Coral Sea, Midway and the Aleutians.* Fort Meade, Md.: National Security Agency, Center for Cryptologic History, 1993.

Pelkonen, Eeva-Liisa, and Donald Albrecht, eds. *Eero Saarinen: Shaping the Future.* New Haven, Conn.: Yale University Press, 2006.

Petzold, Charles. *The Annotated Turing: A Guided Tour through Alan Turing's Historic Paper on Computability and the Turing Machine.* Indianapolis: John Wiley, 2008.

Philofsky, Elliott. "Purple Plague Revisited." *8th Annual Reliability Physics Symposium* (April 1970): 177–185. At tinyurl.com/3dsa9jc (May 2011).

Pickering, G. E., E. G. Mutschler, and G. A. Erickson. "Multicomputer Programming for a Large-Scale Real-Time Data Processing System." *Spring Joint Computer Conference Proceedings* (1964): 445–61.

Piele, Donald T. "Computer Assisted Problem Solving in Mathematics." In *The Computer: Extension of the Human Mind.* Proceedings, Third Annual Summer Conference, College of Education, University of Oregon, Eugene, July 21–23, 1982.

Pine, B. J. "Design, Test, and Validation of the Application System/400 through Early User Involvement." *IBM Systems Journal* 28:3 (1989): 376–85.

Price, Robert M. *The Eye for Innovation: Recognizing Possibilities and Managing the Creative Enterprise.* New Haven, Conn.: Yale University Press, 2005.

Rappaport, Jordan. "Moving to Nice Weather." *Regional Science and Urban Economics* 37:3 (May 2007): 375–98.

Redmond, Kent C., and Thomas M. Smith. *From Whirlwind to MITRE: The R&D Story of the SAGE Air Defense Computer.* Cambridge, Mass.: MIT Press, 2000.

———. *Project Whirlwind: The History of a Pioneer Computer.* Bedford, Mass.: Digital Press, 1980.

Register, Cheri. *Packinghouse Daughter: A Memoir.* St. Paul: Minnesota Historical Society Press, 2000.

Reich, Leonard S. *The Making of American Industrial Research: Science and Business at GE and Bell, 1876–1926.* Cambridge: Cambridge University Press, 1985.

Reilly, Edwin D. *Milestones in Computer Science and Information Technology.* Westport, Conn.: Greenwood Press, 2003.

Reilly, Mark. "Trail of Woe Follows State Fiber Effort." *Minneapolis/St. Paul Business Journal* (March 25, 2001). At tinyrul.com/c2mqy79 (April 2013).

Reynolds, David. *In Command of History: Churchill Fighting and Writing the Second World War.* New York: Random House, 2005.

Rodengen, Jeffrey L. *The Legend of Honeywell.* Ft. Lauderdale, Fla.: Write Stuff Syndicate, 1995.

Rosenblatt, A., and G. Watson. "Concurrent Engineering." *IEEE Spectrum* (July 1991): 22–37.

Safford, Laurance F., and J. N. Wenger. *U.S. Naval Communications Intelligence Activities.* Laguna Hills, Calif.: Aegean Park Press, 1994.

Saxenian, AnnaLee. *Regional Advantage: Culture and Competition in Silicon Valley and Route 128.* Cambridge, Mass.: Harvard University Press, 1994.

Schofield, Hugh. "Minitel: The Rise and Fall of the France-wide Web." *BBC News Magazine* (June 27, 2012). At www.bbc.co.uk/news/magazine-18610692.

Scranton, Philip. *Endless Novelty: Specialty Production and American Industrialization, 1865–1925.* Princeton, N.J.: Princeton University Press, 1997.

Selikson, B., and T. A. Longo. "A Study of Purple Plague and Its Role in Integrated Circuits." *Proceedings of the IEEE* 52:12 (December 1964): 1638–41.

Shurkin, Joel. *Engines of the Mind: The Evolution of the Computer from Mainframes to Microprocessors.* New York: Norton, 1996.

Smiley, Jane. *The Man Who Invented the Computer: The Biography of John Atanasoff, Digital Pioneer.* New York: Doubleday, 2010.

Smith, J. Ernest. "A New Large-Scale Data-Handling System, DATAmatic 1000." *AIEE-IRE '56 Eastern Joint Computer Conference* (December 10–12, 1956): 22–28. At dx.doi.org/10.1145/1455533.1455541.

Smith, Merritt Roe, ed. *Military Enterprise and Technological Change: Perspectives on the American Experience.* Cambridge, Mass.: MIT Press, 1985.

Smith, Robert Elijah. *The Bases of FORTRAN.* Minneapolis: Control Data Institute, 1967.

Snyder, Harlan. "Atlas and the Early Days of Computers." *Bulletin of the National Cryptologic Museum* 4:1 (spring 2001). At tinyurl.com/3btwx9a (March 2011).

Snyder, Samuel S. *History of NSA General-Purpose Electronic Digital Computers.* Fort Meade, Md.: National Security Agency, 1964. At tinyurl.com/265zvjn (June 2011).

———. "Influence of U.S. Cryptologic Organizations on the Digital Computer Industry." *Cryptologic Spectrum* 7:4 and 8:2 (fall 1977 and winter 1978): 65–82.

Soltis, Frank G. *Fortress Rochester: The Inside Story of the IBM iSeries.* Loveland, Colo.: NEWS/400 Books, 2001.

Stern, Nancy. *From ENIAC to UNIVAC: An Appraisal of the Eckert-Mauchly Computers.* Bedford, Mass.: Digital Equipment Corporation, 1981.

Stibitz, George R., and Evelyn Loveday. "The Relay Computers at Bell Labs: Part 1." *Datamation* 13 (April 1967): 35–44.

———. "The Relay Computers at Bell Labs: Part 2." *Datamation* 13 (May 1967): 45–49.

Stix, Gary. "Aging Airways." *Scientific American* (May 1994): 96–104.

Strassmann, Paul A. *The Computers Nobody Wanted: My Years with Xerox.* New Canaan, Conn.: Information Economics Press, 2008.

Takahashi, Shigeru. "The Rise and Fall of Plug-Compatible Mainframes." *IEEE Annals of the History of Computing* 27:1 (2005): 4–16.

Thornton, James E. *Design of a Computer: The Control Data 6600.* Glenview, Ill.: Scott, Foresman, and Company, 1970.

Tomash, Erwin, and Arnold A. Cohen. "The Birth of an ERA: Engineering Research Associates, Inc., 1945–1955." *Annals of the History of Computing* 1:2 (1979): 83–97.

Treadway, James A. *Hard Charger: The Story of the USS Biddle (DLG-34).* New York: iUniverse, 2005.

Van Der Rhoer, Edward. *Deadly Magic: A Personal Account of Communications Intelligence in World War II in the Pacific.* New York: Charles Scribner's Sons, 1978.

Vardalas, John N. *The Computer Revolution in Canada: Building National Technological Competence.* Cambridge, Mass.: MIT Press, 2001.

Watson, Jr., Thomas J. *Father, Son and Co.: My Life at IBM and Beyond.* New York: Bantam Books, 1990.

Wegner, Peter. "A Technique for Counting Ones in a Binary Computer." *Communications of the ACM* 3:5 (May 1960): 322.

Weik, Martin H. *A Fourth Survey of Domestic Electronic Digital Computing Systems.* Aberdeen Proving Ground, Md.: Ballistic Research Laboratories, 1964.

———. *A Survey of Domestic Electronic Digital Computing Systems.* Aberdeen Proving Ground, Md.: Ballistic Research Laboratories, 1955.

———. *A Third Survey of Domestic Electronic Digital Computing Systems.* Aberdeen Proving Ground, Md.: Ballistic Research Laboratories, 1961.

Westwick, Peter J. *Blue Sky Metropolis: The Aerospace Century in Southern California.* Berkeley: University of California Press, 2012.

Wilkes, Maurice V., David J. Wheeler, and Stanley Gill. *The Preparation of Programs for an Electronic Digital Computer.* Cambridge, Mass.: Addison-Wesley Press, 1951.

Willemssen, Joel C. *Year 2000 Computing Crisis: FAA Is Making Progress but Important Challenges Remain.* Washington, D.C.: U.S. GAO, 1999.

Wingerd, Mary Lethert. *North Country: The Making of Minnesota.* Minneapolis: University of Minnesota Press, 2010.

Winterbotham, F. W. *The Ultra Secret.* New York: Harper & Row, 1974.

Wirtzfeld, Roy. "Telex, Inc.: Designed for Profit." *Upper Midwest Investor* 1:8 (December 1961): 6–8, 21.

Worthy, James. *William C. Norris: Portrait of a Maverick.* Cambridge, Mass.: Ballinger, 1987.

Yost, Jeffrey R. "Manufacturing Mainframes: Component Fabrication and Component Procurement at IBM and Sperry Univac, 1960–1975." *History and Technology* 25:3 (September 2009): 219–35.

Zeitlin, Jonathan. "Industrial Districts and Regional Clusters." In Geoffrey Jones and Jonathan Zeitlin, eds., *The Oxford Handbook of Business History*. Oxford/New York: Oxford University Press, 2008, 219–43.

Index

ABC computer. *See* Atanasoff-Berry computer
Advanced Research Projects Agency (ARPA), 99; ARPANET, 122, 137, 147, 261n42
Aiken, Howard, 23, 29, 266n9
air traffic control, 91–93, 257n52
Akerman, John, 140
Albrecht, Bob, 203
Allard, Raymond W., 119, 276
Alliant Techsystems, Inc., 135, 162. *See also* Honeywell
Amdahl Corporation, 114
American Research and Development Corporation (Boston), 12, 108
American Society for Quality Control, 94
American Telephone and Telegraph Corporation, 151, 215, 224. *See also* Bell Telephone Laboratories
American University, 23
Amundson, Neal, 69
Analysts International Corp., 228
ancillary industries, 16, 57, 191, 214, 220, 226–30; definition, 8. *See also* industrial district
Andersen, Elmer (governor), 112
Anderson, Edward, 160

Anderson, Jack, 64
Anderson, Walter L., xi, 276
Antonelli, Kathleen McNulty Mauchly, 34
Apple Computer, 1, 8, 202, 216; Apple II, 209, 211, 215
Arbitron, 121, 134
Arctic Cat, Inc., 3
Aris, Rutherford, 69
Armed Forces Security Agency, 60. *See also* National Security Agency
Army Security Agency, 38. *See also* National Security Agency
ARPA. *See* Advanced Research Projects Agency
ARTS. *See* Automated Radar Tracking System
Atanasoff, John V., 32, 153, 155–59
Atanasoff-Berry computer, 32, 150, 157–59, 265n30
Atari, 210
atomic bomb, 9, 23, 28, 71. *See also* Manhattan Project
Austin, Donald, 214
Automated Radar Tracking System (ARTS), 91–93. *See also* air traffic control

Babbage, Charles, 17, 49, 154. *See also* Charles Babbage Institute
Baker, Eugene L., 260n27, 276
Baltimore & Ohio Railroad, 145–46
Bartik, Jean Jennings, 33, 34, 155
BASIC, 177, 204, 205, 209
Belau, Jane, 214
Bell, C. Gordon, 83
Bell, Gwen, 108
Bellanca–Northern Aircraft, 140
Bell Telephone Laboratories, 18, 35, 39, 149–52, 264n15; Nike missile system, 151; relay computers, 20, 150–53, 264n16
Benchmark Electronics, 222, 234
Bendix Corporation, 121
Bermo (Circle Pines, Minnesota), 224–25, 235
Berry, Clifford, 158. *See also* Atanasoff-Berry computer
"big blue." *See* IBM
BINAC (computer), 3, 36, 59, 150
Binger, James, 160, 161
Bletchley Park, 24, 25, 248n17. *See also* Colossus; cryptography; Enigma
Bloom, H. F., 117
Blue Gene. *See* IBM Blue Gene
Boeing Corporation, 76
Bolt, Beranek and Newman, 122
Bosch Group (Robert Bosch GmbH), 224
Boston, 6, 11, 12, 99, 117, 121, 211. *See also* Massachusetts Institute of Technology; Route 128
Boston Scientific Corporation, 222, 229
Bouza, Erica, 161
Bouza, Tony, 161
BRL. *See* U.S. Army: Ballistic Research Laboratory
Brooks Brothers Lumber Storage, 45
Brown, Gordon R., 276
Brown and Bigelow (publishers), 77
Bruemmer, Bruce, x, 278
Brusentsov, Nikolai P., 54
Burgess, Guy, 61
Burks, Alice, 155, 156
Burks, Arthur, 155, 156

Burroughs Corporation, 153, 168, 190, 195, 213. *See also* Unisys Corporation
Bush, Vannevar, 30, 50, 145, 158
business machine industry, 17, 76, 166–68, 193–98
Butz, Albert, 138, 141

Campaigne, Howard, 23, 55, 249n37, 276
Carey, Patrick, 183, 268n43
Cargill Inc., 2
Carleton College, 211
Carlson, Arne (governor), 213–14
Cedar Engineering. *See* Control Data Corporation
Ceridian Corporation, 134, 211. *See also* Control Data Corporation
Ceruzzi, Paul, 8, 255n31
Charles Babbage Institute, ix, x, 68, 154, 159, 226, 253n59
Chicago, Illinois, 45, 121, 138, 194, 200, 217
Chippewa Falls, Wisconsin, 123, 125, 225
C. H. Robinson Worldwide Inc., 228
Churchill, Winston, 18
Circle Pines (Anoka County), 224
clean rooms, 94, 96
cloud computing, 12, 117, 223, 229
Clover, H. Dick, 247n9, 247n10, 248n16, 249n36, 249n43, 250n10, 252n46
Cohen, Arnold A., x, 49, 54, 55, 60, 67, 276
Cold War, 10, 50, 61, 99, 100, 144, 219
Colossus (proto-computer), 25, 149, 150, 246n14
Columbia University, 131, 165
Communications Supplemental Activity–Washington (CSAW), 22–28, 38–41, 47, 49–54, 59–60. *See also* National Security Agency
Communications Systems Inc. (Hector, Minnesota), 228
Compton, Karl, 108
Computer Assisted Mathematics Programming, 205
Computer Associates, 229, 235
Computer Boulevard (Bloomington), 112–13, 134

Computer Control Company, 145, 149. *See also* Honeywell
computer industry, 1, 5, 7, 14–17, 45, 57, 72–76, 97, 114, 135, 137, 156, 169, 174, 180, 189, 191–98, 203, 213–20, 226–30; employment in Minnesota, 3, 12, 97, 132, 164, 194, 210, 221, 230, 233–43; first digital computers, 150
computer science, 15, 16, 28, 68–69, 79, 99, 145, 215. *See also* information theory
computer services, 111, 116–21, 133, 223, 227–28
Computing Tabulating and Recording Company (C-T-R), 164–66; renamed IBM, 166
Comten Inc., 199, 223, 224, 235
Connecting Minnesota, 217
Control Data Corporation, ix, x, xi, 4, 5, 7, 10–16, 73, 82, 99–134, 163, 174, 180, 198, 203, 206–9, 211–14, 219, 221–22, 224, 229; CDC Cyber (computer), 206–7, 215; CDC 1604 (computer), 14, 100, 109–10, 118, 128, 257n2; CDC 6600 (computer), 14, 101, 110, 121–27; Cedar Engineering, 4, 105–12, 120, 219, 224; C-E-I-R, 121; Commercial Credit Corporation, 121, 133–34; computer services, 116–23, 133–34; Co-op Monitor, 118; Cybernet, 15, 121–22; data centers, 116–20; employment in Minnesota, 131–34, 210, 235; ETA Systems, 211; naming of, 102, 258n7; Northside plant, 129–32; peripheral products, 111–15, 133, 260n27; Storage Module Drive, 114; and supercomputing, 11, 14, 100, 110, 123–29, 133, 211, 214; Supervisory Control of Program Execution (SCOPE), 118
Control Data Institutes, 211
Control Data Systems, 134, 211
Convair, 68, 119
Cray, Seymour, x, 14–15, 276; at Control Data, 100, 104, 106, 110, 120, 258n3; at ERA, 47, 67; and supercomputing, 123–27, 133, 235; at Univac, 81–83

Crichton, Charles, 121, 277
cryptography, 18, 20, 22, 25, 38, 39, 43, 49–51, 54–55, 63, 83, 128–29
CSAW. *See* Communications Supplemental Activity–Washington
C-T-R. *See* Computing Tabulating and Recording Company
Curtiss-Wright Corporation, 18
Cypress Semiconductor, 227

Dartmouth College, 152, 158, 203, 204; Dartmouth Time Sharing System, 204
Data General (minicomputer company), 114, 174, 189
Datalink (Chanhassen, Minnesota), 223
Dataproducts Corporation, ix, 68, 253n59
Davidov, Marv, 159–61, 265n31, 277
Davis, Les, 125
Dayton, Mark, 161, 230
Dayton, Ohio, 9, 11, 17, 26, 39, 43, 141, 223. *See also* National Cash Register Company
Desch, Joseph, 26, 40, 154, 155
differential analyzer, 29–30, 32, 248n23. *See also* Bush, Vannevar
Digi International, 227
Digital Equipment Corporation, 12, 99, 104, 108, 114, 145, 164, 174, 180, 189, 206
Digital River (Minnetonka, Minnesota), 223, 236
Dolan Media, 228
Doriot, Georges, 108
Drake, Willis K. "Bill," 101–8, 277
Driscoll, Agnes Meyer, 22
Duluth, 1–2, 155, 246n7

Eames, Charles and Ray, 171
Eastman Kodak, 39
Eckert, Ernst, 69
Eckert, J. Presper, 31–36, 59, 73–74, 151, 155–57, 277

Eckert-Mauchly Computer Corporation, x, 35–38, 53, 59, 63, 64, 66, 150–57, 248n31. *See also* Univac
EDSAC (computer), 59, 61, 150
EDVAC (computer), 20, 32, 35, 36, 150
Einstein, Albert, 28, 203
Eisenhower, Dwight, 135
Electronic Control Company. *See* Eckert-Mauchly Computer Corporation
Electronic Data Services, 180
Ellison, Larry, 229
Endicott (New York), 166, 168, 170, 173. *See also* IBM
Engineering Research Associates (ERA), ix, 3, 9, 10, 15, 28, 41–42, 47–56, 64, 163, 168, 220–22; Atlas I, 50, 53–61, 64, 150, 251n29; Atlas II, 55, 57, 61, 75–76, 123; Demon, 50, 51, 55; ERA 1101 (computer), 61–67, 75, 117, 252n35; ERA 1103 (Univac Scientific), 62–69, 75–76, 100, 125, 257n2; Goldberg, 50–51, 55. *See also* Univac
Engstrom, Howard, 23, 26, 38–41, 61, 64
ENIAC, 9, 14, 17, 20, 26, 30–38, 58, 150; patent, 37, 137, 150–59, 264n13, 264n17. *See also Honeywell v. Sperry Rand*
Enigma, 25–27

Fairchild, Francis "Dutch," 172
Fairchild Camera and Instrument Corporation, 141
Fairchild Semiconductor, 12, 86, 96, 99, 104, 110. *See also* Intel Corporation
Federal Aviation Administration, 91, 92, 200; Advanced Automation System, 92, 257n50
Federal Reserve Bank of Minneapolis, 8, 194, 200–202
Fein, Louis, 145, 277
financial services industry (Minnesota), 8, 15, 134, 202, 228
Firepond (Mankato, Minnesota), 223, 237
First National Bank of Boston, 145–46
First National Bank of Minneapolis, 102, 107

First National Bank of St. Paul, 41, 49
Flanders, Ralph, 108
Fletcher, Abbot L., 104
Fletcher, Fremont, 104, 108
Flint, Charles, 166
Flynn, Timothy, 214
Folz, Bernice, 214
Ford Motor Company, 41, 77; Twin Cities Assembly Plant, 189
Forrest, Henry S., 106–9, 277
Forrestal, James, 39, 63
Forrester, Jay, 79, 80. *See also* SAGE; Whirlwind computer
Fort Snelling, 77
FORTRAN, 177, 204
Fry, Thornton, 102
FSI International Inc., 227
Fujitsu, 114
Furey, Tom, 176, 177

gender: and employment, 4, 30, 33, 77–79, 194–97
General Dynamics Corp., ix, 10, 76, 222, 237
General Electric Corp., 35, 66, 102, 113, 135, 147, 151, 153, 204, 253n51
General Fabrication, 198–99, 237
General Mills Inc., 2, 3, 5, 117, 191
Georgia Institute of Technology, 117
Gerstner, Louis, 15, 180
global enterprises, 220, 223. *See also* industrial district
Gödel, Kurt, 28
Goldberg, Max, 214
Goldstine, Adele, 34
Goldstine, Herman H., 32, 34, 155
Gopher Ordnance Works, 18
Government Code and Cypher School. *See* Bletchley Park
Grabbe, Dimitry, 126, 277
Gray, George, 66
Green Giant Company, 2
Groupe Bull, 135, 149
Groves, Leslie (general), 66, 71
Gunderson, Richard C., 117, 277

Hagley Museum and Library, 154, 277
Halley, Woods, 160
Hamline University, 45
Hamming, Richard, 129, 151
Hanke, Jay, 228
Hansen, Pete, 176
Harris, James D., 106, 120, 277
Harvard Mark I (computer), 23, 29, 150, 168. *See also* IBM
Harvard University, 23, 29, 52, 53, 69, 108
Hedger, Dick, 277
Heller, Walter, 201
Hendrickson, Arnold, 60, 88
Hewlett-Packard Company, 8, 180, 206
HickoryTech Corp., 228
High-Speed Computing Devices, 53
Hill, John L. "Jack," 43, 55–61, 74, 83, 277
Hiss, Alger, 61
Hitachi, 114; Hitachi Global Storage Technologies, 227
Hoerni, Jean, 86
Holberton, Frances E., 34, 38, 155, 249n35, 277
Holberton, John, 34, 155
Hollerith, Herman, 165–68. *See also* IBM
Honeywell, 3–8, 10, 12, 15, 63, 89, 113, 237; C-1 automatic pilot, 141–44; Datamatic 1000, 145–46; Honeywell Aerospace, 117; Honeywell Information Systems, 149, 226; Honeywell International, 135; kitchen computer, 146–47; merger with Allied Signal, 135; thermostat, 15, 136–39
Honeywell, Mark, 138
Honeywell Project, 135, 159–62
Honeywell v. Sperry Rand, x, 14, 17, 137, 149–59, 247n20. *See also* ENIAC: patent
Hopper, Grace, 23, 29
Hormel Foods Corporation, 2
Hurwicz, Leonid, 69, 201
Hutchinson Technology, 8, 14, 223–26, 237

IBM, 22, 23, 26, 39, 166; ASC supercomputer, 174; Automatic Sequence Controlled Calculator, 168; changing logo, 167; employment in Minnesota, 164, 221, 238, 268n35; Endicott, 174; Engineering and Technology Services, 180, 186, 187, 223; Fort Knox, 175; Global Services Division, 180; nickname as "big blue," 171, 172; OS/2 personal computer, 178; Project SAGE, 80; Rochester, x, xi, 5, 10, 12, 15, 163, 169–87; San Jose, 164, 174; Selective Sequence Electronic Calculator, 168; Silverlake project, 175, 178; System/3, 175, 177; System/36, 175–77; System/38, 175–77; System/360, 114, 123, 148, 172, 174, 175; Thomas J. Watson Research Center, 171, 180; Watson supercomputer, 14, 164, 187
IBM AS/400, 174–79, 183, 186
IBM Blue Gene, xi, 14, 163–64, 179–86, 268n38
IBM 1401, 173, 177
IBM Selectric typewriter, 171
IBM 701 "Defense Calculator," 75, 168
IBM 7030 Stretch, 14
IBM 650, 54, 58, 63, 68, 168
Illinois Scientific Developments, Inc., 154
Imation, 14, 220, 223, 227, 237
industrial district, 6–8, 12, 16, 137, 138, 164, 172, 173, 180, 189–91, 219, 224, 230. *See also* ancillary industries; global enterprises; integrated anchor; specialist auxiliary
information services. *See* computer services
information theory, 18, 129, 151. *See also* computer science
Institute for Advanced Study (Princeton, New Jersey), 28, 53, 155, 203
integrated anchor, 7, 219, 221–23. *See also* industrial district
Intel Corporation, 1, 8, 100, 180, 181
International Association of Machinists, 77
International Business Machines. *See* IBM
Internet, 99, 122, 129, 137, 180, 200, 215, 217, 228; invention of "surfing," 215; service providers, 216, 217
Internet Alley, 8

Internet Gopher, 16, 215–17, 270n32
Iowa State University, 32, 153, 157–58

Japan: attack on Pearl Harbor, 21–22; Imperial Japanese Navy, 23–24
Jawer, Bruce, 179
Jennings, Betty Jean. *See* Bartik, Jean Jennings
Jeopardy! (quiz show), 14, 164, 187
Jobs, Steve, 209
Johns Hopkins University, 31, 155
Joseph, Earl, 214

Kamp, Thomas G., 106, 111–15, 277. *See also* Control Data Corporation
Kehrberg, Kent, 209
Kemeny, John G., 203, 204, 277
Kershaw, Jay, 47, 86
Keye, William, 60, 106
Keys, Ancel, 18
King, Martin Luther, Jr., 131
Kisch, Robert, 106
Kramer, Alwin, 21–22
K-ration, 18

LaFrenz, Dale E., 203, 270n14, 277
LARC. *See* Univac: Livermore Advanced Research Computer
Larson, Earl (judge), 155–58
Lawn-Boy, 2
Lawrence Berkeley Laboratory, 214
Lawrence Livermore National Laboratory, 10, 14, 73, 101, 110, 119, 127, 164, 181–86, 200
Lawson Software (St. Paul), 8, 229, 238
Leas, Vernon, 88
Lécuyer, Christophe, 8
Ledbetter, Carl, 214–15
LeVander, Harold (governor), 206–7
Lewis, Sinclair, 2
Lewis, Steve, xi, 180–83, 186, 277
Lilly, Richard, 41, 49
Lincoln, Neil, 125, 276
Lipscomb, William N., 69
Lockheed Corporation, 76, 118, 207

Lockheed Martin Corporation, xi, 10, 213, 222, 241
Lorenz teleprinter cipher, 25, 247n14
Los Alamos National Laboratory, 14, 28, 33, 128, 204

MacArthur, General Douglas, 21, 63, 71–72, 254n3
Mack, Mike, 222
Maclean, Donald, 61
Madison, Wisconsin, 169–70
Malcolm Baldrige National Quality Award, 174, 179
Mall of America, 112, 132, 227
Manchester Baby (computer), 150
Manhattan Project, 18, 28, 29, 71, 80. *See also* atomic bomb
Mankato State University, 148
Manning, E. J. "Jim," 105. *See also* Control Data Corporation: Cedar Engineering
Marshall, Alfred, 6. *See also* industrial district
Marshall, General George, 21
Massachusetts Institute of Technology (MIT), 18, 26, 35, 53, 80, 99, 135, 169, 204. *See also* Whirlwind computer
Mathiowetz, Curt, xi, 183–86, 277
Mattel, Inc., 211
Mauchly, John W., 30, 32, 38, 52, 151, 155–59. *See also* Eckert-Mauchly Computer Corporation
Mauchly, Kathleen McNulty. *See* Antonelli, Kathleen McNulty Mauchly
Mauchly, Mary, 30
Mayo Clinic (Rochester, Minnesota), 169
McCahill, Mark P., x, 16, 215–17; and POPmail, 215. *See also* Internet Gopher
McCarthy, John (computer scientist), 204, 277
McCarthy, Joseph (senator), 38, 64
McData Services, 227
McDonald, Robert E., 76, 277
McNally, Irvin, 82–83
Meader, Ralph, 26, 64

MECC. *See* Minnesota Educational Computing Consortium
medical device industry, 8, 16, 138, 177, 180, 219–29; directory, 228
Medtronic, Inc., 211, 219–23, 228–29, 271n1
Melman, Seymour, 160
Meltzer, Marlyn Wescoff, 34
Memorex, 8, 226
Metalcraft Machine and Engineering, 225, 239
metropolitan Minneapolis–St. Paul area, 5, 105, 191–98, 217, 230
Metropolitan Stadium, 112
MICE. *See* Midwest Internet Cooperative Exchange
Michigan Hospital Service/Blue Cross, 145
Microelectronics and Computer Technology Corporation, 133
Microsoft, 1, 10, 180, 186, 209, 216, 223, 239
Midway industrial district, 45–47, 52, 57, 224
Midwest Internet Cooperative Exchange (MICE), 217, 228
Midwest Technical Development Corporation, 4, 12, 108, 259n21. *See also* venture capital
Miles, James, 106
military-industrial complex, 135
minicomputers, 99, 114, 137, 145, 149, 174, 180. *See also* Data General; Digital Equipment Corporation; Hewlett-Packard Company
Minneapolis, 1, 12, 100, 104, 121, 129, 132–38, 159, 161, 200, 223; industrial district, 137–39, 199, 225; Internet exchange, 217; schools, 205. *See also* Federal Reserve Bank of Minneapolis
Minneapolis Athletic Club, 102
Minneapolis–Honeywell Regulator Company, 3, 12, 105, 137, 138. *See also* Honeywell
Minneapolis–Moline Company, 2
Minnesota: computer manufacturers (1972), 199; early history, 1, 77; state fair, 4, 10; state university system, 207; support of computing, 13, 206–15; weather, 5, 246n7. *See also* computer industry; metropolitan Minneapolis–St. Paul area; University of Minnesota
Minnesota Department of Education, 207
Minnesota Directory of Manufacturers, 220, 224, 229, 233–43
Minnesota Educational Computing Consortium (MECC), xi, 133, 149, 197, 200, 202, 207–14
Minnesota Historical Society, 15, 77, 137, 160, 207
Minnesota Mining and Manufacturing (3M), 3, 8, 12–14, 56–57, 77, 89, 133, 203, 206, 214, 220, 223
Minnesota Multiphasic Personality Inventory, 69
Minnesota Precision Manufacturing Association, 222
Minnesota's Medical Alley, 191, 219. *See also* medical device industry
Minnesota Supercomputing Institute, 13
Minnesota Transfer Railway Company, 45
Mississippi River, 2, 77
MIT. *See* Massachusetts Institute of Technology
MITRE Corporation, 80
Mohawk Aircraft Corporation, 140
Montgomery Ward, 46
Moore, Gordon, 100; Moore's Law, 73, 93, 181
Moos, Malcolm: and "military industrial complex," 135
Moscow State University, 54
Motorola, 93, 96, 216, 223
Mullaney, Frank C., 47, 55–58, 60, 66, 100, 102, 106, 115, 123, 277
Multiband Corp., 228
Mumma, Robert E., 27, 40, 154, 277
Museum of Modern Art (New York), 171

nanotechnology, 230–31, 272n23
National Aeronautics and Space Administration (NASA), 3, 62, 99, 152

National Bureau of Standards (NBS), 3, 35, 38, 49, 53; SEAC computer, 3, 49, 53, 150; SWAC computer, 49
National Cash Register Company, 17, 26, 39, 154, 165, 168, 223
National Center for Atmospheric Research (Boulder), 183
National Security Agency (NSA), 10, 14, 39, 49–59, 61, 127–29, 155. *See also* Communications Supplemental Activity–Washington
National Semiconductor Corporation, 109
National Urban League, 131
Naval Computing Machine Laboratory, 26, 43, 47, 50, 83, 250n13. *See also* Dayton, Ohio; Engineering Research Associates
Naval Tactical Data System (NTDS), 79–90, 200, 256n46; and Seymour Cray, 83; programming, 89–91; reliability and failure analysis, 93–96
Nebraska Avenue Complex (Washington, D.C.), 23–27, 52, 60, 61
NEC Earth Simulator (computer), 184, 186
Net Perceptions (internet company), 239
New York Stock Exchange (NYSE), 7, 40, 101
Nier, Alfred O. C., 55, 67
Nimitz, Admiral Chester W., 41
Nintendo, 186
Norberg, Arthur L., x–xi, 67, 276–77
Norden bombsight, 141–43, 263n4
Norris, William C., x, xi, 10, 277; at Control Data, 100–104, 106–9, 113, 120–23, 129–33; at CSAW, 23, 28, 38–39; at ERA–Univac, 41, 61, 64, 73–77
Nortech Systems, 227, 236
Northern States Power, 117
Northrop Aircraft Company, 36, 59
Northrop Grumman Corporation, 213
Northwest Airlines, 40, 49, 140
Northwestern Aeronautical Corporation (NAC), 40–42, 47, 64
Norwalk, Connecticut, 66, 71, 75, 108. *See also* Remington Rand Univac

Noyce, Robert, 100
Noyes, Eliot, 171
NSA. *See* National Security Agency
NSFNET, 217
NTDS. *See* Naval Tactical Data System
NVE Corp., 230, 239

Occupational Outlook, 194, 196, 200
Occupational Wage Survey, 193
office machines. *See* business machine industry
Office of Naval Research, 35, 53. *See also* U.S. Navy
Office of Strategic Services, 64
Open Systems (Shakopee, Minnesota), 223, 240
OP-20-G. *See* Communications Supplemental Activity–Washington
Oracle Corporation, 229, 240
"Oregon Trail" (computer game), 149, 202, 211. *See also* Minnesota Educational Computing Consortium
Osterwise, Larry, 179

Parker, John E., 40–41, 63–64, 71, 140, 163, 277. *See also* Engineering Research Associates
Paske, Tom, 173
Patterson, John, 165. *See also* National Cash Register Company
Patton, Peter, 229, 278
Pearl Harbor, 9, 20–22, 141
Pearson, Drew, 63–64
Pendergrass, James T., 35, 39, 52–55, 61, 278
People's Computer Center, 203
People's Computer Company, 203
Perlman, Lawrence, 211
Perot, H. Ross, 180
Philadelphia, Pennsylvania, ix, 8, 11, 17, 32, 35, 76
Pike, Zebulon, 77
Pillsbury, Charlie, 161
Pillsbury Company, 2, 191, 205
Piper Jaffray, 102

PLATO (computer system), 15, 133, 211, 240, 263n66. *See also* Control Data Corporation
Polaris Industries Inc., 3, 225
Polaris submarines, 110
Polar Semiconductor, 227
Poughkeepsie, New York, 168, 173. *See also* IBM
Prescott, Edward C., 201
Price, Robert M., ix–xi, 108, 119–21, 211, 253n59, 278
Prime Computer, 114
Princeton, New Jersey, ix, 28, 203. *See also* Institute for Advanced Study
purple plague, 94

radar, 9, 18, 21; in air traffic control, 91; as model for computing, 55, 67, 82, 145; in NTDS, 83–90; in SAGE, 80
Rajchman, Jan, 155
Ramsey Engineering, 4, 74. *See also* Hill, John L. "Jack"
Rand, James, 66, 67, 71–75. *See also* Sperry Rand Univac
Rand, Paul, 171, 172
RAND Corporation, 38, 204
Raytheon, 114, 145; Datamatic 1000, 145–46; RAYDAC computer, 145. *See also* Bush, Vannevar
RCA Corporation, 114, 135, 151, 153
reduced instruction set computing (RISC), 85, 125–26, 255n33. *See also* Cray, Seymour
Redwood Falls, Minnesota, 132, 134, 198
Remington Rand Univac, 3, 17, 36, 38, 64–67, 71–72, 168. *See also* Sperry Rand Univac
RISC. *See* reduced instruction set computing
Riverside Machining and Engineering (Chippewa Falls, Wisconsin), 225
Robinson, Herbert W., 121, 278
Rochester, Minnesota, 1, 12, 163–64, 169–72, 179, 198, 222, 224. *See also* IBM
Ronning, Joel, 223. *See also* Digital River
Roosevelt, Franklin D., 21, 168

Rosemount Engineering, 4, 112
Rosenberg, Julius and Ethel, 61
Route 128 (Boston), 6, 12, 99, 189, 246n17
Rowan, Tom, 123
Rubens, Sidney M., 56, 278
Ryden, Arnold J., 101, 102, 107, 108, 278

Saarinen, Eero (architect), 171. *See also* IBM: Rochester
Safford, Laurance, 22
SAGE, 79–89. *See also* IBM; Whirlwind computer
Sargent, Thomas J., 201
Schleicher, David L., 176, 278
Schumacher, Mike, 119, 278
SCOPE. *See* Control Data Corporation: Supervisory Control of Program Execution
Scranton, Philip, 7–8, 219–20
Seagate Technology, 13, 134, 241, 273n22
semiconductor industry, 73, 93–99, 181, 200, 226, 227, 230; purple plague, 94
Service Bureau Corporation, 15, 116, 123. *See also* Control Data Corporation; IBM
Shannon, Claude, 8, 18, 20, 151
Shepherd, William G., 68, 107, 259n16. *See also* University of Minnesota
Sherman Antitrust Act, 154, 156
shopping malls, 112, 191. *See also* Mall of America
Sibley, Henry, 77
Silicon Graphics, 133, 235, 241
Silicon Valley (California), 5–12, 100, 189, 200, 219, 227, 229, 246n17
Sims, Christopher A., 201
Sirica, John (judge), 154
Smaby, Greg, 214
Smith, Robert "Doc," 203
Snelling Shops. *See* Twin City Rapid Transit Company
Snyder, Harlan, 60
Soltis, Frank G., 164
Sony, 186
Southdale Mall, 112, 191
Spangle, Clarence, 149, 226

specialist auxiliary, 7, 220–25. *See also* industrial district
Spence, Frances Bilas, 34
Sperry Defense Products, 213
Sperry Flight Systems, 213
Sperry Marine, 213
Sperry Rand Univac, ix, xi, 5, 10–11, 16, 67, 72–76, 92, 95, 100–108, 133, 137, 152–59, 213, 214
spin-offs, 4, 11, 14, 72, 99, 105, 112, 145, 206, 220, 227–28; defined, 4
spintronics, 230
SPS Commerce Inc., 228
Sputnik, 99, 100. *See also* Cold War
Standard Oil of California, 118, 119
startup companies, 3, 11, 47, 53, 73, 99, 100, 108, 133, 163, 198, 219
Stein, Marvin L., ix, xi, 63, 68, 69, 278. *See also* University of Minnesota
Stephens, Francis, 214
Stibitz, George, 151, 152, 158, 264n16
St. Jude Medical Inc., 222, 229
St. Louis Park Historical Society, 265n31
stored-program computer, 3, 32, 36, 54, 55, 59, 149–51, 168
St. Paul, 1, 2, 9, 13. *See also* metropolitan Minneapolis–St. Paul area; Midway industrial district
St. Paul Radio Club, 55
St. Paul Union Stockyards, 45
Strickland, Ed, 106
supercomputing, 1, 11–15, 54, 110–14, 123–29, 179–87, 214, 225, 245n1, 268n43
Svendsen, Edward, 82, 83
Svendsen, Mike, xi, 97, 257n57
Sweatt, William R., 138, 139

Tabulating Machine Company, 165. *See also* Computing Tabulating and Recording Company
Tata Consultancy Services, 223, 228, 230
Teitelbaum, Ruth Lichterman, 34, 155
Telex Communication, 109, 224, 242
Texas Instruments, 96, 227
Thelen, Ed, 83, 129

Thornton, James, 47, 110, 125
3M. *See* Minnesota Mining and Manufacturing
TIES. *See* Total Information for Education Systems
time-sharing computers, 99, 116, 118, 121, 135, 137, 148, 204, 207, 209
Tomash, Erwin, ix, x, 47, 63, 68, 75, 253n59
Tompkins, Charles B., 53
Tonka Toys, Inc., 2
Toro Manufacturing Corporation, 2, 41
Total Information for Education Systems (TIES), 205–6. *See also* Minnesota Educational Computing Consortium
Trousdale, Elmer B., 123, 278
Turing, Alan, 8, 18, 25–28, 151
Twin Cities. *See* metropolitan Minneapolis–St. Paul area
Twin City Rapid Transit Company, 46; Snelling Shops, 46, 57, 224

Ulam, Stanisław, 128
Unisys Corporation, xi, 213–14, 190, 221–22, 229, 241
Univac: antenna coupler, 76–78; Athena (missile guidance computer), 83–84, 88; Livermore Advanced Research Computer, 54, 73; Semiconductor Control Facility, 95–97. *See also* Sperry Rand Univac
Univac File (computer), 91–92
Univac I (computer), 36, 53, 59, 60, 74, 150, 251n31
Univac 1100 (computer), 207
Univac Park (Eagan), 77, 89, 97
Univac II (computer), 53, 73–76
University of Michigan, 155
University of Minnesota, ix, 2–5, 12–19, 23, 47, 55, 63, 67–68, 82, 107, 112, 125, 133, 135, 140, 155, 160, 161, 183, 200–203, 207, 214–17, 228, 229, 231; Digital Technology Center, 215
University of Pennsylvania, ix, 17, 30, 154; Moore School of Electrical Engineering, 30, 34–36, 51–59

University of St. Thomas, 214
Upper Midwest Investor, 5
U.S. Air Force, 36, 38, 76, 80, 81, 91, 141, 251n31
U.S. Army, ix; Ballistic Research Laboratory (BRL), 17, 30, 152
U.S. Bureau of Labor Statistics, 193, 194
U.S. Census Bureau, 36, 59, 165
U.S. Department of Defense, 63, 99, 162
U.S. Department of Energy, 214
U.S. Department of Homeland Security, 23
U.S. Department of Labor, 131
U.S. District Court, 154
U.S. Justice Department, 123
U.S. Navy: Bureau of Ships, 49, 54, 57, 88; Naval Electronics Laboratory, 82; Naval Ordnance Laboratory, 56; Naval Postgraduate School, 82, 107, 109–10; Office of Naval Research, 35, 53; Women Accepted for Volunteer Emergency Service (WAVES), 26–28. *See also* Naval Tactical Data System
U.S. Office of Technology Assessment, 209
U.S. Patent Office, 36, 153, 154, 163, 165
U.S. Post Office, 110
U.S. Securities and Exchange Commission (SEC), 102
U.S. Treasury Department, 145

vector processing, 125, 128. *See also* supercomputing
Venona project, 61. *See also* National Security Agency
venture capital, 4, 8, 12, 99, 108–9, 211, 214. *See also* Midwest Technical Development Corporation
Vietnam War, 91, 159–60
Von Neumann, John, 8, 9, 28–32, 51, 128, 155, 156

Wallace, Neil, 201
Walsh, Joe, 49
Wassenberg, D. P., 104
Watson, Thomas J., Jr., 127, 169, 171
Watson, Thomas J., Sr., 165–69
WAVES. *See* U.S. Navy
Weaver, Warren, 155
Wells Fargo & Company, 136, 138, 140
Wenger, Joseph, 22, 23, 39, 41, 49, 61, 64
Whirlwind computer, 56, 79, 254n22. *See also* Massachusetts Institute of Technology; SAGE
Wilkes, Maurice, 53, 59
Williams tubes, 57, 62
World War II, xi, 1, 3, 17, 18, 20. *See also* Pearl Harbor
World Wide Web (WWW), 16, 180, 213–17, 223
Wosniak, Stephen, 209
Wright Field (Dayton, Ohio), 141

Xerox Corporation, 180, 216, 227, 243

Yale University, 23
Yorktown Heights, 163, 164, 171, 179, 182, 183. *See also* IBM: Thomas J. Watson Research Center
Yost, Jeffrey, x, 257n57
Young, Whitney, 131

Zacharias, Jerrold, 82. *See also* Massachusetts Institute of Technology
Zemlin, Richard A. "Dick," 118–19, 278
Zuse, Konrad, 149; Zuse Z3 (computer), 150
Zworykin, Vladimir, 155

THOMAS J. MISA is Engineering Research Associates Land-Grant Chair in History of Technology in the Program for History of Science and Technology and professor in the Department of Electrical and Computer Engineering as well as director of the Charles Babbage Institute at the University of Minnesota. He has written or edited nine books, including *Gender Codes: Why Women Are Leaving Computing* and *Leonardo to the Internet: Technology and Culture from the Renaissance to the Present*.